W9-BRD-749

DISCARD
PORTER COUNTY
LIBRARY SYSTEM

# HUNDREDS OF INTERLACED FINGERS

# HUNDREDS OF INTERLACED FINGERS

## A KIDNEY DOCTOR'S SEARCH FOR THE PERFECT MATCH

Vanessa Grubbs, MD



Amistad

Some names and identifying details have been changed to protect the privacy of patients and medical professionals.



HarperCollins books may be purchased for educational, business, or sales promotional use. For information, please e-mail the Special Markets Department at SPsales@harpercollins.com.

FIRST EDITION

*Designed by Michelle Crowe*

Library of Congress Cataloging-in-Publication Data has been applied for.

ISBN 978-0-06-241817-3

17 18 19 20 21 LSC 10 9 8 7 6 5 4 3 2 1

*For My Robert*

# CONTENTS

PART IV:
*You're Gonna Miss Her When She's Gone*



PART V:
*Letting Her Go*

# PART I

## IF HE ONLY HAD HER

# 1

## SOMEDAY COMES

Twenty-six-year-old Robert Phillips made an appointment to see his primary care doctor. He didn't feel well and he hadn't felt well for a couple of months. He was exhausted. When he wasn't in a work meeting or working out, he was sleeping. He blamed all the travel he had been doing for work lately. He had a headache that he couldn't shake. Maybe the stress was getting to him. There was a strange metallic taste in his mouth that reminded him of the time he was five years old and tasted rebar when the men were working on Big Mama's house. His stomach, back, legs, and hands cramped when he wasn't walking around. Maybe he was just really dehydrated, he thought. He had been working out a lot, yet he was gaining weight.

He had been peeing more often too, getting up several times a night to go to the bathroom. Maybe that's why he was so tired.

"Your chest doesn't hurt?" asked the physician assistant as she looked up wide-eyed from the blood pressure gauge at Robert. She was used to seeing patients with high blood pressure—in the range of 150 to 180 over 90 to 110—in the clinic. They struggled with her recommendation for a low-salt diet and didn't like taking the one or two blood pressure medicines she had prescribed. But now she stared at 220 over 120. She had never seen such high numbers before. It must have been hard for his heart to continue pumping blood through his body when it was meeting with that much resistance. His chest *had* to hurt.

"No, but my head really hurts," Robert answered. His head hurt so bad he didn't care that he was seeing this person instead of his primary care doctor, Debra Daniel. He liked Dr. Daniel. She was a tall, pretty Black woman, maybe in her mid-forties he thought. She was cool, like a big sister. She would always see him within two days. And with his busy and erratic schedule, he often needed an appointment within a short window of time. It had been a year since he had seen her, but she had been his primary care doctor since he aged out of Dr. Santoyo's pediatrics practice. He didn't know who this plain-looking thirty-something-year-old White woman was. He had never seen her before.

"I'm going to take your urine and blood and I'll call you with the results. Go home and wait for a call," she said.

Two hours later, just as Robert was bringing a slice of the Round Table pizza he had ordered to his mouth, his phone rang. It was 1 p.m. and he was starving, but he dropped the slice to answer the phone. It was Dr. Daniel.

"I made an appointment for you at the ER. I need you to go right now. Do you need an ambulance? Do you have a car? Can you go right now? I need you to go right now," she said without pause or even a breath.

"OK," he said, having never heard this level of urgency in her voice before. He hung up the phone and immediately headed to the ER, his pizza forgotten. The only thought he could hold in his pulsating head was *What is going on?*

Dr. Daniel hadn't mentioned his kidneys, but after getting more blood tests the ER doctor asked Robert who his nephrologist was. He was staring at the ceiling from his gurney, still wondering what was going on, when Matty Kravitz walked up. Kravitz had been his nephrologist for nearly nine years, when his first kidney doctor, Barry Gorman, retired from his San Pablo, California, clinic practice and transferred Robert to Kravitz's care.

As much as Robert liked Gorman he disliked Kravitz. Gorman wanted to see Robert every three months. He wanted him to get blood and urine tests before the appointments. At the appointments Gorman seemed genuinely interested in him as a person. He reviewed the test results ahead of time and explained them.

Kravitz was content to see Robert once a year and ordered tests at the visits, then never called with the results. And even worse, Kravitz never remembered Robert from visit to visit. This day was no different.

"How did you know to ask for me?" Kravitz asked. The bags beneath his eyes and the beak nose were the same, but the crescent moon his receding hairline left behind had widened and he had lost what Robert guessed to be about fifty pounds since he saw him the previous year.

A wave of heat seared through Robert's chest. "Because you *saw* me," he responded, incensed. "How much weight did you lose?" Robert asked, in hopes of jogging Kravitz's memory. He needed his doctor to remember him.

"Oh, I guess you did see me." Kravitz nodded slowly, his eyebrows lifted and mouth turned down, conceding the point.

Robert wanted to punch him. *You're lucky I'm cramping,* he thought.

"Do you know what happened to you?" Kravitz asked as he flipped through the pages of Robert's ER chart.

"No. What happened to me?"

"Your kidneys are failing."

Now Robert felt like he was being punched. Ten years ago he was warned this day would come. He just didn't think it would be this soon. And he didn't know this is what it would feel like.

"We've got to put you on the dialysis machine," Kravitz went on. "We're going to put a catheter in your chest. Later we'll look into getting you a fistula, an access in your arm."

*Catheter? Fistula? Access?*

All of these words were foreign to Robert. He had never heard of any of them before.

But he did know what a dialysis machine was. He remembered the words he said when he was sixteen years old and walked past a dialysis unit with his mother on his way to Dr. Gorman's office the first time—*If I gotta do that, you might as well kill me.* It wasn't hypothetical anymore, at some vague point in the distant future that might not even come. It was now. And it was irreversible. "Do this or die," he was being told.

Hours later he woke up to find himself in the hospital's acute dialysis unit tethered to a dialysis machine for the first time. The room had six beds, three against each wall. Each bed had to its left a dialysis machine that with its tentacle-like tubing looked like a five-by-two-foot monster to him. One other patient, Dr. Kravitz, and two nurses were there.

He thought he would feel the push and pull of his blood through the tubing and the wastes leaving his body, but he didn't. His chest ached from where the catheter had been tunneled under his skin, but the soreness would be gone in a day or two, he was promised. Dialysis wasn't as bad as he thought it would be.

Again he stared at the ceiling, this time wondering what was to come. Just yesterday he was dreaming about the next steps in his budding career in politics and wondering if the next woman he dated would be the one he'd make his wife.

Now he wondered if he would have to move back in with his parents. How he would be able to do this dialysis thing and keep up his work pace. If any woman would even want him now that he had this catheter hanging out of his chest. Now that he was sick.

*Maybe I won't have to do this for long,* he hoped.

# 2

## ULTERIOR MOTIVES

It was 2003 and the culminating meeting for the Alameda County Medical Center (ACMC) strategic planning exercise. I was invited to participate because after my internal medicine residency at ACMC's Highland Hospital in Oakland, California, I stayed on as faculty. As an attending physician, I was working to build a meaningful program to address issues of diversity in the medical center. I called it the Office of Diversity Affairs (ODA). My vision was that ODA would recruit more doctors of color to mirror the mostly Black and Brown patient population we cared for, train the staff and doctors already there to be more receptive to learning about cultures not their own, and expand interpreter services to more effectively engage the dozens of languages spoken within our walls. An effort of this magnitude needed the support of the hospital leadership, and I had the CEO's backing. I wanted the support of each member of the board of trustees too, and I made it my mission to meet each one.

Robert Phillips was a trustee.

The meeting was soon to begin, so when I saw Robert walking across the room, I headed toward him. He was younger than the other trustees. He wore a dress shirt, no tie, and slacks and appeared to be in his early thirties like me. He was tall and had the whole impeccably groomed goatee and mustache with a shaved head look going. He had a linebacker's build and a belly that looked like a small keg. My "ideal" was a slender build with a six-pack, and what was worse, he had fair skin and green eyes, which were major affronts to my deep color issues that dated back to childhood. I blame my father and his side of the family for them.

My family is like a box of crayons with only shades of brown, from sandy tan to almost black, but nearly everyone at the family reunions on my father's side was in the earth tone to almost black range. One family reunion in particular stood out. I couldn't have been more than six or seven. My five siblings all seemed to be vying for a legitimate excuse to be spared the ten-hour ride from Spring Lake, North Carolina, to Dothan, Alabama, near the Florida border, but only two were successful: Milton, sixteen years my senior, who had joined the Army right after high school, and Cynthia, who had some school event that could not be missed. The rest of us piled into the station wagon along with our bags and a cooler full of sandwiches, fried chicken, even pineapple cake, because stops were only for buying gas and peeing. I was probably the only one who didn't mind the hours in the car. As the baby of the family—the surprise baby, *not* the accident, as my siblings often teased—I was allowed to spread out across everyone in the backseat (seat belt laws and booster seats did not yet exist). Michael, at eight years my senior, was the closest in age to me. But from Michael to Milton, my siblings were like stair steps, eighteen to twenty-four months apart.

I remember standing with Janet, my oldest sister, and Regina, the next-to-the-youngest girl, in the park on the day of the reunion. I looked up as the women gathered around in groups of three or four at a time to dote on Janet.

"*Ooh,* you so pretty!" they said over and over again. Janet smiled and giggled thank you again and again. Regina stood silently. No one was talking to her.

I had always thought of Regina as the prettiest of us girls. I didn't follow why they weren't telling Regina how pretty *she* was, with her smooth skin and petite features rather than the not-so-petite nose and pimples that Janet and Cynthia had and that I would soon develop. But I soon realized that Regina's mahogany skin made her less pretty in their eyes.

Back home, a similar theme was playing out. My father didn't like the TV shows with Black people. *Good Times, The Jeffersons,* even *The Cosby Show.* When one of them was on, he grunted his disapproval and often made us change the channel. And, to my mother's chagrin, my father fawned over his brother's biracial wife as if she were the most beautiful thing on the planet. As if he wanted her for himself. If only Uncle Roosevelt would just hurry up and drink himself to death already.

I resented the notion that fairer skin was prettier or better. I railed against it. Even in myself. When my first cousin teased, "Why you got blond hairs on your arms?" I took to lying in the swinging bench in the backyard for a few hot summer days in an attempt to get darker. My baby dolls had to be brown. Only Baby Alive and Holly Hobbie got a pass because one pooped and the other came with an oven that baked cakes with a lightbulb. And as I got older, as rebellious as the nerdy child I always was could be, never even *considering* stunts like sneaking in and out of the house late at night, I resisted in the men I chose. Enter my first husband, almost black. Robert was the opposite. I would *not* acknowledge his attractiveness.

"Hello, I'm Dr. Burt," I said, my married name at the time. "I lead the Office of Diversity Affairs." I offered my hand for a shake and smiled innocently. Innocent, wide-eyed smiles. I gave these smiles instinctively to men to avert or avoid sending any hint that there would ever be some kind of sexual energy between us.

"Hello, Dr. Burt," Robert returned, shaking my hand. He tilted his head slightly to the right as he blinked once slowly and gave me a strained closed-lip smile that communicated *Now what do you want from me?* but said nothing else. No *Ah yes, I've heard about your efforts* or *Oh, I'd like to hear more about that,* as the others had.

Not expecting this response, I didn't quite know what to

say, so I nodded a smile and walked away, feeling that he was blowing me off.

The next time I saw Robert was a few months later—in a mini-documentary titled *Worlds Apart,* a four-part series on how culture impacts health care. I used the ten-minute video featuring Robert as part of an ODA training I organized for the Highland internal medicine faculty. I hadn't put it together that the Robert in the video was the Robert I had met. In the video, Robert spoke about the two and a half years of his life on dialysis. It didn't occur to me to question why Robert was on dialysis. Rather, my full attention was captured by what he was saying in the video. He spoke about the mistrust of doctors among the Black community. About how it seemed like the Whites in his dialysis unit either were going to get a transplant or already had one, while the Blacks were in the dialysis unit for the long haul. About how the transplant center staff made him feel like he was no more than an irritating file they had to pull out every time he called to ask about when he might get a transplant. "We'll call you," they would say.

Afterward, my assistant connected the dots for me. Robert video was Robert trustee. "I thought he had nice eyes," she added, and I rolled mine in response. My father would think his eyes were nice too. But not me. I would not be impressed. He wasn't *that* engaging in the video. And he only *barely* made me smile in spite of myself.

Not long after the training, it was announced that he and several other trustees resigned abruptly in protest of certain decisions made by county leadership, and Robert faded into the depths of my memory.

Hello, it's Dr. Burt," I answered brightly, still typing on my computer keyboard. My office was an eight-by-eight-foot box with furniture that was too big for it. It was located about a mile and a half away from my colleagues' offices in the department of medicine suite, but it had a window and a separate space for the quarter-time assistant ODA could afford. It was mine and I was happy to have it.

"Uh, hello, this is kind of awkward, but," an unfamiliar female voice returned, "are you married to Joe Burt?"

I stopped typing. "Yes."

"I've been dating him for the past several months."

I looked at the phone, as if *it* had betrayed me. "What?"

She went on to provide a few details of her involvement with my husband. Things he had told her. Facts he had reshaped. It felt like someone had just hit the replay button on my first few months with him seven years prior, with just the names of the son and estranged wife he claimed changed. The rest of the script was the same.

I knew this mistress wasn't his first, but it was hard to catch my breath. My peripheral vision had left me, so I stared between my knees at the floor beneath my office chair. The dark blue carpet tiles appeared to be receding like crashed waves on a beach.

And then it dawned on me—*This is my chance to be free. Because this one called me.* She had given me an airtight reason to leave my marriage that no one (except his mama) could fault me for taking.

I almost thanked her.

I hung up the phone gently and turned to the window. Sunlight streaked across my expressionless face as tears streamed down. Splats on my blouse roused me like a snap out of a trance. I wiped my face with the backs of my hands and pushed the buttons to call Melanie Tervalon's home.

Melanie was in her early fifties and a huge, international, phrase-coining figure in diversity work. Her medical school class valedictorian speech calling out institutional racism in medicine made her a legend in my mind. And she was my friend.

In my first efforts to launch ODA, I reached out to her for help. I was naive and clueless, and she was generous with her time and advice. Melanie had a regality about her when she walked, her shoulder-length blond-silver hair and chin tilted up just so. But she would also not hesitate to tell it like it is, with a certain realness in her voice. I loved that she shared this part of herself with me very early on in my knowing her. It helped me see that the part of me that felt at ease with the janitorial staff was as important as the part that might someday command the most elite of audiences like she did. Both were strengths. Both were parts to be embraced.

As a very junior person in my work, I felt lucky just to know someone of Melanie's stature, let alone call her a personal friend. But because of Melanie's reputation for frankness, I soon learned that dropping her name to gain some advantage could backfire. People had strong, usually extreme reactions to Melanie—adoration or disdain and nothing in between—and you could never predict which. But that was work. This was personal.

Melanie answered her phone on the second ring. Minutes later I sat with her at her kitchen table. I wanted her to tell me what to do. But as free-flowing as Melanie was with her advice for work, she was tight-lipped when it came to personal matters. Instead she offered her house keys and a futon to me and my three-year-old baby for as long as it took me to figure it out.

Nearly a year later, I found an unfamiliar male voice on the other end of my phone. It was Robert.

Melanie was his friend too and had encouraged him to call me to get the skinny on a few of the doctors at Highland. He was still on dialysis, for more than four years by then, and was trying to elicit the support of physician advocates in his effort to get a kidney transplant.

I spoke frankly about my colleagues at Highland as Melanie told him I would, and he chuckled as I spoke. The conversation ended with his smooth baritone voice inviting me to a dinner at Melanie's home to discuss his health with a larger group.

"Of course. A friend of Melanie's is a friend of mine," I said.

I thought the dinner would be for a group of seven or eight, but it turned out to be just Robert, Melanie, and me. It was an early June evening in Oakland, which meant we were still wearing light jackets. Melanie's home was a quaint Craftsman she had raised her children in. After her divorce many years before, she had painted every room a different vibrant color.

I arrived first. Having come straight from work, I was wearing black slacks and a long-sleeved polyester-blend blouse. I was standing in Melanie's throwback 1950s kitchen when Robert walked in cradling a loaf of French bread and some fancy cheeses in his arms. As he looked at me, head to toe and back up again, his left brow lifted and the edges of his mouth pressed downward in a silent *Aww shit*, like he was seeing me—and my curves—for the first time. I instinctively crossed my arms over my chest and smiled a stilted hello. It was too late for one of my wide, innocent smiles. A very sexual energy was already in the air. *Shit,* I thought.

The three of us sat down to dinner. Conversation flowed easily and bubbled with laughter as we ate the meal Melanie prepared. I reminded him of how he blew me off at that first encounter. He denied it. That was not his intention, he said. He was just used to people coming up asking him for things.

"Whatever. You were rude." I rolled my eyes but smiled in spite of myself, betraying my forgiveness. He smiled back.

I was so distracted by Robert, I didn't notice what Melanie was doing with her face, which was why I was so surprised when after we had barely put our forks down, she stood up, grabbed her purse, and announced, "I'm going to run to the store and get us some ice cream for dessert. I'll be back shortly." And with a twirl into her coat she was out the door. I watched the door close with eyes wide and jaw dropped in *what the hell just happened* wonder. Melanie had left Robert and me there sitting across from each other at her little kitchen table, our knees almost touching.

We looked at each other. A moment of awkwardness gave way to close-lipped smiles. He leaned forward, his arms folded on the table.

"So what's your story?" I asked. Still licking the bitter wounds that the lies the last man told had left behind, I expected little more than bullshit to come from the lips of this man in front of me.

Instead he spoke candidly about where he was in his life. Perhaps too candidly, like there was no time to waste, but I followed suit. I told him about my three-year-old son and divorce under way. He told me that he lived with his parents. *What kind of grown man tells a woman he just met that he lives with his mama?* I tried not to change my facial expression, but he read it anyway. One who passed out from very low blood pressure, and no longer felt safe living alone, he offered. He was putting his cards on the table as if to say *This is who I am. Your move.* There was an honesty in his eyes that was foreign to me. I liked it. It didn't even matter anymore that they were green.

Later when we walked to our cars, we caught each other looking back over our shoulders, like some clichéd scene from damn near every boy-meets-girl movie. We smiled.

Melanie!" I hissed into the phone as I drove home. "Did you set that up?"

"I did if it works out," she said.

# 3

## BID WHIST

The next day I received flowers at work. A beautiful bouquet of orange baby calla lilies in a shallow, dark green ceramic dish. The card read: *My apologies if I greeted you with anything less than a smile when we first met. In earnest, Robert Phillips.*

*Who says "in earnest"?* I thought, but smiled. Scenes from the night before flashed through my mind—lingering gazes, involuntary smiles, looks over shoulders—and sent a flood of warm tingles through me. I smiled some more. Even twirled one of my locks around my finger. I took the bait and dialed his number. I planned to thank him for the flowers and ask him out to lunch.

He answered on the second ring. His voice sounded thick

like he was just waking up . . . or didn't feel well. He cleared his throat.

"Oh . . . hi, it's Vanessa. I'm sorry, did I wake you?" Though it was early afternoon.

"No, I'm at dialysis," he said plainly and cleared his throat again.

"Oh! I'm sorry!" As if I had just walked in on him naked. "I can call you back another time."

I was pulling the phone away from my ear when I heard him say, "No, it's fine. Really. I can talk."

"OK," I said, though I still felt uneasy. I had been snapped out of girly girl into a vision of his reality as a person on dialysis. But soon conversation was easy again.

"I'd love to," he said to my lunch invitation. "I'm free tomorrow."

It was a Thursday. We met at Kincaid's in Jack London Square, a surf-and-turf restaurant overlooking the Oakland Bay. I wore a scarf wrapped around my head so that my locks spilled out of the top. It was my in your face rejection of a European standard of beauty and projection of the Nubian princess that I am or at least feel like on the inside hairdo. Robert wore slacks and a button-down shirt and a slight approving smile on his face. Neither of us planned to go back to the office that day, so lunch was leisurely—and heavily laden with alcohol, which no doubt contributed at least in part to the easy conversation. It felt like we were the only patrons there. I was aware only of him, the view of the bay, and the waiter appearing occasionally to ask if we wanted another round.

By my third Grey Goose Appletini and his third double Jack on the rocks, I had probably shared too much about life with my ex and he had shared things I hadn't even thought to ask about yet. While I knew that only a functioning liver was needed to process alcohol, I didn't know how much he'd be paying for taking in all that extra liquid—the nausea, vomiting, and maybe even shortness of breath—until it could be dialyzed off because his kidneys couldn't pee it out. So for the time being, it was all smiles and lingering gazes. And lots of laughter.

Afterward, we walked together to the parking garage.

"I had a really good time," he said. "We should go out again soon."

I smiled, nodding in agreement. "I'm free this weekend," I said, and immediately wished I could suck back in the words that had just fallen out of my face as I saw Robert's brows lift in surprise and his mouth spread into a closed-lip smile.

I flushed and then tried to explain myself. "I mean . . . It's just that I alternate weekends with my son's father . . . Avery's with him this weekend . . ." I stammered. Still a bit too eager, but at least I was letting him know that my eagerness was laced with some logic and that my time with my son would not be interrupted. But truth be told, waiting more than a couple of days to see Robert again felt unacceptably far away.

He smiled at my awkwardness. "OK. I'll call you later and we can make plans for this weekend."

He opened his arms tentatively, his face questioning if I would welcome a hug. I took a step toward him and reached my arms around his neck. As his arms wrapped around me, I instinctively laid my head on his chest, but at least I didn't hang on too long. At least I don't think I did.

I stepped back and bashfully brought my eyes to his. I found them twinkling. We said our good-byes and I smiled all the way home.

He called the next day.

One might think that, as a doctor, I might be particularly hesitant to get involved with a man I knew had end-stage kidney disease because any illness with "end-stage" in its name is by definition life-threatening. But it was more of myself as a woman going through a divorce that questioned my ability to pick a good man. So naturally, I presented the male specimen to a small panel of trusted girlfriends and asked what they thought I should do.

First there was Phyllis.

"*Mira.* Look it," she said to me, revving up to make her point. "Who do you think you are?"

This was not the reaction I expected. I just looked at her, my eyes wide in surprise.

Phyllis was a statuesque Latina and the oldest girl in a family of eight kids where there was never enough money but usually too much alcohol. She managed to escape her upbringing and become an ICU nurse, then after a decade went to medical school because she was tired of following doctors' orders. She wanted to be the one giving the orders. She was a year behind me in training, my intern when I was a resident at Highland. And she was my sister. Secretly, we even referred to ourselves as SISS—Sistas in Stirring up Shit, for it was the two of us who started giving voice to the issues that led to the formation of ODA. It was Phyllis who drove down to Anaheim with me for an Internal Medicine Board Review course and was content to live off peanut butter sandwiches for lunch and dinner because we were that broke. And it was Phyllis who I called at 2 a.m. when my ex's incessant phone calls and antics in those early weeks of our separation had me second-guessing my decision to leave. Then it was, "*Mira.* Look it, you have to be strong, Vanessa!" as if she had me by the shoulders, shaking me through the phone line. "*Mira.* Look it" always preceded some you may not want to hear this but goddamnit you're gonna tough love, Phyllis-style.

"Who do you think you are to disregard somebody just because he is sick?" she continued without blinking. "Who died and made you so perfect?"

I was speechless as I mentally noted one for Robert.

Melva was next.

"Girl, please," she said with a wave of her hand.

Melva was a petite psychiatrist with a Diana Ross and the Supremes bob and was on her way to becoming the Dr. Phil of Baltimore. We met through the Kellogg Scholars in Health Disparities program, which I entered that year to get a taste of what a research career in medicine might be like. I figured she would be able to give me advice *and* diagnose the crazy at play.

"He just needs a kidney and he's good, right?" As if I were perseverating unnecessarily.

I smiled as I noted two for Robert.

It was everyday dolphin earring–wearing Jane—because her last name was Dolphin—the older nurse I worked with at Highland and trusted to babysit my son, who said, "Don't get involved with a sick man. You'll be taking care of him later." Her words sent a defensive jolt through my heart that told me what it knew all along. I wanted to take the chance with Robert. No need to ask anyone else.

It was on the second date that I decided for sure that I wanted to move forward with Robert. We were sitting at a restaurant bar waiting for our table when Robert confided that he didn't have too many second dates. He laid down his cards, faceup, on first dates. Most decided the game he presented was too complicated. Like Bid Whist when all they were up for was a game of Go Fish.

He seemed nervous, unable to sit still. Like *That first date was just the first hand, girl.* I was nervous too. The entirety of my renal training at Highland was a monthlong rotation in my second and third years of residency, so my knowledge of the nuances of the kidney was minimal. Sure, I knew that normally we are born with two kidneys and that they are located in the back, one on either side of the spine behind the membrane sheet that wraps around other organs in the abdominal cavity, and are only partially protected by the last two ribs. I knew a few things that could make the kidneys stop working—like high blood pressure and diabetes and certain poisons and rare blood vessel diseases—but most of the details and inner workings of the kidneys were a mystery to me. And since the bulk of my training was in the hospital, I mostly saw dialysis patients in trouble—usually because of problems with their hemodialysis access.

I understood hemodialysis on only the most basic level. I knew the dialysis machine is like a big kidney, siphoning away extra fluid, waste, and electrolytes like potassium from the blood because the patient's kidneys can no longer pee these things out. I knew the "access" is literally how we gain access to the patient's blood. Through the hemodialysis access, the patient is connected to the dialysis machine so that their blood can be routed through it at a rate of about a cup and a half a minute. Hemodialysis access is a lifeline because without it—a failed access—there is no way to connect to the machine. Not being able to connect to the machine means no dialysis. No dialysis means all the extra fluid, waste, and electrolytes like potassium are left behind and accumulating to the point that soon there would be no life. If I was going to risk my heart on Robert, it was important to me that Robert didn't have problems with his access.

"Do you have a fistula?" I asked.

"Yeah," he said excitedly, as if he was happy that he didn't have to be the one to bring it up. "Wanna see it?"

"Yes," I said without hesitation. I needed to see it.

In other areas of medicine, a fistula is not a good thing. In other areas, a fistula is an abnormal connection between body parts, like the obstetric kind that forms when in a very young mother's narrow pelvis the baby's skull presses the vagina against her intestine or bladder for days during labor. Eventually the pressure leaves a hole behind, and the woman is left constantly leaking stool or urine. But when it comes to dialysis, an arteriovenous fistula is created intentionally when the surgeon sews two blood vessels in the arm together—an artery to a vein.

I imagined the body's system of blood vessels as an intricate map of roadways. Normally arteries branch off the heart like a main thoroughfare that splits off into four-lane boulevards, then two-lane avenues, then one-way streets carrying oxygen-rich blood to all the body parts. At the end of one-way streets are alleys—capillaries. But instead of dead-ending, the alleys connect to an almost identical set of streets heading back to the heart—the veins. The streets meet the highways meet the main vein thoroughfare carrying oxygen-drained blood to begin again the heart's cycle of pumping it back to the lungs for more oxygen. The surgeon's connection of an artery and vein to create an ateriovenous fistula bypasses the narrowest streets and alleys altogether so that traffic no longer has to slow down for red lights or stop signs—blood flow increases from about half a cup a minute to two cups a minute as soon as the stitches are in place and to somewhere between another cup and two faster just weeks later. This fast-paced blood flow forces the vein to become longer and thicker. Within a couple of months it will be "mature"—long enough and thick enough to withstand two dialysis needles too large for a regular vein puncturing it at least three centimeters (a bit more than an inch) apart three times a week and with a blood flow potentially faster than other types of access yet slow enough that a person won't bleed to death in seconds—as an artery would—when the needles are removed. The fistula is the best type of dialysis access even though there can be complications: blood sometimes clots in them; part of the blood vessel wall can become dangerously thin and at risk of rupturing; or part of it could narrow, restricting blood flow. But it is still best because there is nothing foreign left behind to potentially create additional problems, as with the arteriovenous graft or the hemodialysis catheter, the two other types of access.

The arteriovenous graft is next best. It lays in a loop beneath the skin in the arm after the surgeon sews one end of the man-made version of a blood vessel to an artery and the other end to a vein. Surgeons do this when the patient's vein is too narrow to ever mature. The graft is more prone to infection and clotting than the fistula, but it is ready to withstand dialysis needles within a couple of weeks.

The catheter is the worst of the choices. Yes, it is convenient, ready to use immediately after being inserted into one of the body's large veins in the chest or groin, and patients like that its tails hang outside the body connected to the dialysis machine. No needles necessary. However, the pencil-thick plastic tube is often the source of dangerous blood infections and clotting, much more frequently than the graft. Infection and clotting are often followed by replacement. Each replacement catheter damages the blood vessel a bit more, each insult sending the body into repair mode, laying down a bit more scab until, soon, no catheter can traverse—which is a problem because there are only four places to put a catheter without having to go through the spine or liver first.

And having to go through the spine or liver first is never a good sign.

Robert hurriedly rolled up his left sleeve to reveal the curved, thickened blood vessel coursing up his forearm like a snake. I could feel him studying my face for its reaction. Many of his past dates' faces had contorted to match the deformity they perceived on his arm. But worse were the ones that dissolved into pity. All of them had snatched their fingers away as if the snake would turn on them. This was his test.

I ran my hand up the length of it, feeling the strong skin and buzz, buzz, buzz pulsating beneath my fingertips. No thinned areas that were at risk of bursting open, taking his life in minutes. No faint stirring of blood signaling that all flow would soon cease to exist. His lifeline. *Oh, yes,* I thought, *we can do this, boy.* I sensed him exhale as he watched my face spread into a smile.

Nevertheless, I still felt nervous anticipation around him, especially on our fifth date when he still hadn't kissed me. He called me on it.

"Why you seem so sheepish?" he asked at the end of our date as we sat together in his car in the parking lot where mine was parked.

"I don't know . . . I mean . . . I don't know." I blushed.

He smiled knowingly.

"Just lean toward me," he said softly but confidently in his baritone voice. Instinctively I did as I was told. He leaned toward me.

"And kiss me," he said.

My heart pounded hard in my chest, but his lips were soft and he gently tickled my tongue with his between our just slightly more than church wedding–appropriate open-mouthed lip motions. I tingled.

"We had a moment," he said with a smile after our faces separated. A new flourish of tingles swept through me. I was thinking the same thing.

I exhaled, feeling like I was in a romantic movie. We said our goodnights and I tried not to twirl back to my car.

But not too many more dates later he said something that I thought might make the desire to twirl stop forever: "It ain't like it's kidney failure."

It was a refrain Robert would often repeat to himself and others to remind them there were worse things in life—like what he dealt with in the day-to-day—compared to whatever it was that was upsetting them in the moment. Those words sparked our first fight.

I forget what I was fretting over that prompted him to say it. I said nothing in the moment because I am one of those unfortunate souls who are slow to snap back. I wish I were one of those people who can go toe-to-toe with anyone. I want to be able to freestyle rap or play the dozens effectively. *Yo mama so . . .* Wait, I need time to think.

My brain needed to decipher why it was that my heart thumped, my face heated, and that wave of prickly tingles washed over me when he spoke those words. So taken by surprise, my brain needed hours to accept and apply meaning to this intrusion.

My opportunity to speak my hurt came just hours later, when Robert asked me how I was feeling about something that was bothering me. We were on the phone and I was driving, Bluetooth earpiece curled around my right ear. Hands gripping the steering wheel at ten and two and leaning forward, like a little old lady trying to see over the wheel. Robert was at dialysis.

"Doesn't really matter," I said flatly, "it's not like it's kidney failure." *Take that*, I thought.

Now he was the one to pause, but just for a beat because he is a toe-to-toer. "What is that supposed to mean?"

By then I could articulate my thoughts. But so could he. Our voices raised. Mine about how hurtful it felt for him to dismiss what I was going through because it didn't hang me in the balance between life and death as did his kidney failure. His about how what I was so worried about *was* small compared to being in a dialysis chair.

I could picture him where he always sat for dialysis since the incessant work calls he would take when he dialyzed in the large open space of the dialysis unit disturbed the more than two dozen others who just wanted to sleep through their three or so hours on the machine, which won him a permanent assignment to the room off to the side of the dialysis unit like some sort of a posh private suite, when really it is where patients with diseases like hepatitis and tuberculosis typically would be sent so they could be isolated from the other patients. Still, his voice was so loud that Glen, the technician monitoring him, peeked in to ask, "Who you talking to like that?"

I imagined a flock of birds, sensing a devastating storm was on its way, flying off in a wave.

Instead, a peculiar thing happened. He heard me. I listened to him. We both had had our share of experiences in which a disagreement so small nobody remembers the details days later had devolved into name-calling and phone slamming. This was different. This was good. This was how it was supposed to be. It was then that I knew that Robert and I were good for each other. That we were good together.

And so we began to intertwine ourselves like grown folks. Grown folks have been through enough to know not to waste time looking for perfection because we can see our own flaws clearly. The goal was finding a partner who was as crazy as we were and willing to put up with our crazy. We knew not to waste time waiting at least three days before calling or pretending to be busy on a Friday night. Or even being afraid to say I love you first. Robert and I were on the phone when he said it that first time.

"It's kind of hard to say"—his voice began to tremble—"but I am starting to feel love for you."

"Why is that so hard to say?" I said.

"Uh . . . I don't know," he stammered and changed the subject.

It was some time after we hung up before I realized that *feeling love for me* was Robert's roundabout way of telling me he loved me for the very first time and, without meaning to, I had essentially responded with "*That's nice. Thank you*" in return.

*Idiot!* I thought of myself, because, oh, how I did love him back.

Robert was like no other. Quickly he won me over, no doubt in part because he was the opposite of what I was used to. He had *volunteered* the not-so-pretty details of his life on the first date. I was impressed. He would drive the thirty-six miles through traffic from his home in Richmond to see me in suburbia often and without hesitation. I was wooed. He could rattle off a brief history of labor unions and politics in America. I was titillated.

It wasn't long before Robert invited me to visit with him while he dialyzed. Having seen patients on dialysis during my training, the yards of tubing filled with blood and the faint disinfectant smell of the dialysis unit did not faze me. But during training, visits to patients were quick assessments of dialysis orders and vital signs in a small hospital room with a capacity of only four patients. At that time, I was the internal medicine resident physician two years out of medical school and just one year away from being able to become a medical board–certified internist. I was usually sleep-deprived and always eager to get on to the next patient. The clipboard of a patient's vital signs and doctor's orders was the only detail I really recall.

My visits to Robert were as girlfriend. As girlfriend, just weeks into being in love, there was nowhere else I wanted to be.

The outpatient dialysis unit was huge compared to the hospital dialysis unit. Like bees dotting from flower to flower and bringing nectar back to the hive, dialysis nurses and technicians in pajama-like scrubs buzzed around the thirty recliners against the walls to a central nurses' station. Each dialysis chair had a side arm that swung a TV in front of the patient's face. Most of the patients were asleep. With diabetes and blood vessel disease and kidney failure running together as they often do, it was not surprising that some patients were missing a part of a leg or two.

"You're six kilos up," Glen, the technician, said to Robert. "You have to not drink so much between runs, man."

Robert stepped off the scale, not bothering to respond this time. He felt thirsty all the time. He wished he could pee some of the extra fluid off his body. He no longer even felt an urge to pee. He missed that.

He sat in the burgundy recliner freshly wiped down with disinfectant after the last patient. He laid his left arm palm up on the armrest as he wished for nothing outside of the norm this day. His new normal. It had taken him three years to get here, to a point where he could dare wish for things to go smoothly. He'd learned who he could trust to stick him. He'd learned that when he traveled to never allow a nurse to stick him. The nurses had gone to school longer, but the techs stuck fistulas all day. He wanted a tech. He'd learned how to guide a new tech, what angle their needle should take. The wrong angle wouldn't get inside the fistula, but rather between skin and fistula. It was a stick that hurt more, but he wouldn't really know it was in the wrong place until the machine started to return his blood back to him and the arm swelled immediately, bringing with it a sudden sharp pain that gave way to throbbing. But what was worse was that a needle in the wrong place meant no dialysis that day. They would give him a dose of chalky Kayexalate to drink, and it would give him the diarrhea necessary to get his potassium down to a level where there was no worry that it would rise high enough to stop his heart. But he would still be left with the nausea and the heaviness and shortness of breath from too much waste and too much fluid left behind. All because some non-fistula-sticking nurse had messed up. He remembered how he had to resist the urge to punch with his right fist when it happened. *Bitch!* he once shouted instead.

Glen, he could trust. Needle setup in hand, Glen rolled over the short black cushioned stool and took a seat. He pulled up the recliner's side table until it clicked in place and laid down his supplies. He tore off a strip of white tape and touched one end to the edge of the table. It hung there like a windless flag. He hung three more.

The needle looked like it could hold pieces of wood together. Instead it was attached to eight inches of clear tubing. Facing Robert, he held the needle at a forty-five-degree angle over the fistula traveling up Robert's left forearm. He noted the two sets of marching dots from previous sticks. He aimed for a space a quarter inch above the last. First the arterial needle that would pull the blood from his body into the machine. He secured the needle in place with two strips of white tape. He repeated this for the venous needle that would return the blood to his body. Robert hardly flinched. He was used to this violation. After all, this was better than having the catheter like in the first couple of years. With the catheter, he was always in fear of sepsis, infection coursing through his blood. He remembered the shivering and sweating it produced.

Glen connected the machine tubing to the needle tubing.

For four hours, Robert had to be careful not to tug at the clear tubing tether from the dialysis machine to the two large-bore needles spearing his fistula. Newspapers, books, cell phone, DVD player, and TV were close at hand to help pass the time. He remembered how once, on a work trip, when he had to go to a different dialysis center, he had an allergic reaction to the dialysis machine filter, the foot-long plastic cylinder filled with fibers that served as the artificial kidney. He recalled the immediate sneezing, eye watering, chest tightness, difficulty breathing—and then the extra forty-five minutes he had to sit *after* his dialysis treatment with the appropriate filter, waiting for the Benadryl he needed to treat the allergic reaction in the first place to wear off. He had learned to pack his own filters. One for every two days of travel.

He felt himself lightening as the extra fluid he had accumulated since his last run on the machine was being siphoned away. Too quickly? He worried that his punishment for drinking too much between runs would be light-headedness and cramping in his legs, his back, his jaw, his hands. His everywhere. Again. He ignored the hum of the blood pressure cuff tightening around his right arm every fifteen minutes until the end of his dialysis treatment, his "run," when it had his full attention and he willed the device to display a reading high enough for him to be able to stand up and walk out of the place, only to start over again in two days. This was a good dialysis day. Only low blood pressure and fatigue. This was to be expected.

He learned a long time ago to ease back into those dialysis-free hours. Trying to do too much too quickly landed him on the floor with the panic-stricken faces of his colleagues hovering above him. His routine had become going to the movie theater right after dialysis, allowing the air-conditioned darkness and story unfolding on the screen to soothe and distract him long enough for the post-dialysis exhaustion to subside.

He supplemented the movies with a mani-pedi every other week. Few would have imagined Robert, who was a linebacker in college, to be in the nail shop with such regularity. I folded into his pattern, but even I had assumed that Robert would much rather have been outside hiking, running, and jumping under our sunny California sky; that this stillness was forced upon him by illness. But truth was, Robert refused camping trips because for him "roughing it" meant to stay at a hotel without a spa.

The petite Vietnamese ladies always flocked around him as if male energy was the one thing the shop had been craving.

"You want cut short?" the one at his toenails would ask. The one at his hand was sure to startle, pulling her hand back when she happened upon his buzzing fistula. "What happened to you?" she would ask.

"I'm sick," he would reduce it down to, hoping that would be enough of an explanation. It still embarrassed him. He rarely wore a short-sleeved shirt because of it. It could be eighty degrees out, but T-shirts were only for home. People would stare. The bold few would ask. Their own embarrassment, or not wanting to seem nosy, or not wanting to make Robert even more uncomfortable, forced them to silence that inner four-year-old's mantra: "But why?"

*Because my kidneys stopped working.*

*But why?*

*Because. Because they said I have something called . . . because I don't know why.* He'd change course, not feeling like making his tongue roll out the strange combination of syllables that was his diagnosis.

*But why?*

*Because they don't know either.*

Next was to fold Robert into my life. He would be the first man Avery would meet since I left his father. I invited him to meet us at Chuck E. Cheese's, Avery's favorite hangout. My tummy fluttered with anxiety. Just the week before, as Avery and I played on the monkey bars near our apartment, he had announced, "I don't like boyfriend," when I told him Mommy had one.

"Why not?" I responded at the time. "Daddy has girlfriends and you like them."

"Daddy needs a girlfriend."

"Mommy doesn't need a boyfriend?"

"No."

"What does Mommy need?"

"Me!" He grinned.

But by the end of the afternoon, after playing games and eating bad pizza, Avery said good-bye to Robert with a hug.

"You're a nice man, Robert," he said, with his head resting on Robert's belly and his little arms stretched around.

Robert smiled with surprise and placed his left hand on Avery's back while his right hand patted his head.

I smiled too. Robert had passed *my* test.

Which was why at the end of that summer when Robert and I were sitting at a little outdoor café table enjoying some frozen yogurt I asked, "So what you trying to do?" like some slang version of a father asking *"And what are your intentions with our daughter?"* Though my own father would have been much more likely to wait on the front porch with his shotgun barrel resting menacingly on his right shoulder if a boy hadn't brought his daughter home by the expected time.

"I'm looking for a wife," Robert announced, his right eyebrow twitching and lips quivering.

He was nervous. *How endearing.* I smiled to myself, but said, with a hint of attitude, "What does that mean? Just any woman will do? Just add Vanessa and stir?"

"No, not just any woman . . ." he pushed back. "It ain't like you a prize."

I inhaled sharply and looked at him with my eyebrows knitted. *What the hell is that supposed to mean?*

He read my face and quickly tried to clean up his words. "I mean, you have a child *and* a crazy ex-husband. . . . It's about you."

"Whatever, sucka. I am a prize," I said, returning to my frozen yogurt and not hearing him ask me what it was I was trying to do with him.

# 4

## THE OTHER ROBERT PHILLIPS

The following spring I lay in a Bay Area preoperative care unit hospital bed, though as a doctor I was used to standing nearby as someone else lay in one. My tummy somersaulted with so much nervous excitement that it bubbled out into my throat and wouldn't allow me to take a deep breath. The sweatpants, sweatshirt, bra, panties, socks, and sneakers I wore there were tucked neatly into a large yellow plastic bag labeled PATIENT BELONGINGS. In their place was only the hospital-issued light blue cotton gown with too few ties in the back to make my booty feel securely hidden and tan booties with the no-slip treads. My locks were stuffed under a periwinkle bouffant bonnet that didn't quite match the gown.

It was Thursday, April 14, 2005, and what I hoped was

the day after the last day of dialysis that Robert would have to endure since his kidneys failed some six years prior. I was about to give him one of mine.

Although it had been months after we started dating before it ever crossed my mind that I could be the source of his new kidney, I always believed things would work out in his favor, that he would get a kidney transplant. I *had* to believe in order to allow my heart to soften for him in the first place. To let him in. But I *knew* things would work out from the moment I heard about The Other Robert Phillips.

It was My Robert, as I had become accustomed to referring to him, as there was often more than one Robert in the room or conversation, who first told me about The Other.

"People in the dialysis unit who had already had a transplant but ended up back on dialysis kept saying that having a kidney transplant was even worse than dialysis. That you go through all that surgery and all those pills and they only last a couple of years," he told me. "Even the nephrologists gave me conflicting information, so I wanted to find out for myself," he said, and he began to research. First kidney transplant. Average kidney transplant survival. Longest kidney transplant survival.

It wasn't long before he found him in the archived profiles of the American Association of Kidney Patients: Robert Phillips—the longest surviving kidney transplant recipient. In 1963—a decade before My Robert was even born—The Other Robert Phillips, a truck driver from Virginia, received a kidney transplant from his sister. According to the article, the transplant had been thriving forty-one years and counting and they weren't even compatible ABO blood types. He was type A, which can usually only receive from type Os and other type As. She was type B, which usually can only give to type ABs and other Bs. This was all the proof My Robert needed that a kidney transplant held the promise of a life much better than dialysis could ever offer.

Seeing his own name in print gave Robert an extra glimmer of hope that he too could have this promise fulfilled. *This could be me,* he thought. But he had to be careful. *Could this be me?* He couldn't hope too much. Hoping too much was dangerous. Hoping too much made getting through the day-to-day too hard. The focus was getting through *this* day, this moment. And because his "moment" as a person on dialysis could last another five, ten, fifteen, twenty years, to treat it like it was a short-term thing meant to risk survival—of body *and* spirit.

People who didn't learn this lesson could not survive, Robert believed. The spirit would go first. Then the body. Like it did with Monsor.

"Monsor sat in the dialysis chair next to me," Robert began the story. "I remember him saying over and over again, 'I ain't gon' be here long with y'all. I ain't gon' be here long.' He was sure a transplant would happen for him quickly."

He went on wistfully, looking somewhere over my head to the right, smiling in the memory. "Monsor was tall and muscular in the beginning. You wouldn't know he was on dialysis. His dress, his attitude, the way he carried himself. He would have his dialysis run and be on about his business. . . . He was older, dark-skinned, and wore his hair pressed straight so that it came down almost to his shoulders. He always wore jewelry. Bracelets. He had a hoop ring in his nose and an earring that were connected on the left side by a gold chain studded with diamonds. He always wore silk shirts and walked into the dialysis unit bare-chested with his shirts open. And he wore these round glasses with a black frame across the top that flared out like horn-rimmed glasses."

"Hold on," I interrupted. "Was Monsor a pimp?"

"Yeah, he was," he said boyishly, as if surprised his description gave it away. "But pimpin' wasn't paying enough so he had a nine-to-five too. About three years into it and a transplant didn't happen, he was like 'Fuck it,' and became difficult to be around. Angry. He didn't eat right. He lost his job and a lot of weight, but he was too proud to apply for assistance. His dress changed. He went from being a clotheshorse to Dumpster diving for stuff to put on. He stopped getting his hair done. The stereotypical pimp fried, dyed, and laid to the side hair became an unkempt Afro. Eventually they found him dead on the street." His smile faded as he looked down at his hands.

"Did that make you sad?" I asked.

He instinctively shook his head. "He wasn't the first person who died," he said with a shrug, but there was the twitching at his right temple that I had learned to expect whenever feelings came up for Robert.

Robert was much better at expressing his feelings in writing. Over the years he often had flowers delivered to my office or home just because he was thinking of me and sent me e-mail love letters and random texts with love quotes. In person, however, feelings were things that he braced himself for until they passed, like earthquakes, hoping there wouldn't be too much collateral damage left in their wake. He had the same approach with my feelings. He usually shifted his eyes left and right to avoid watching tears stream from my eyes. Sometimes he would just look at me blankly as if I was a stranger and he wanted to inform me that water was coming out of my eyes, like he would let someone know they had spinach in their teeth. At best he offered what felt like a bear paw heavily patting my knee: *There, there, it's gonna be OK.* My Robert is romantic. From a distance.

For me the story of The Other Robert Phillips was a sign. A clear, indisputable sign that My Robert would get a transplant and that it would last decades upon decades too. No matter that The Other Robert Phillips was White and shared DNA with his donor and had a completely different cause of kidney failure than did My Robert. It was enough for me to hold on to.

I went with My Robert to his kidney transplant evaluation appointment in 2004. It was a beautiful late November day in San Francisco. The fog had not yet rolled in over the Golden Gate Bridge. Robert and I tingled with anticipation. He had been told that after five years of being on the kidney transplant waiting list, he was finally near the top and that the visit was a mere formality to make sure he was still healthy enough to get a kidney transplant and still wanted one. And just like on TV and in the movies where people who get CPR simply get up and walk out of the hospital, they had Robert believing he would be assigned a pager that would beep us out of sugarplum dreams in just a few days.

And despite the MD behind my name, I actually imagined him going from clinic to hospital gown to get his new kidney the next morning.

Instead, we sat in a clinic exam room listening to a series of people whose job it seemed was to talk Robert out of even wanting a transplant. They filed in and out, one after the other.

"Even after your insurance pays," said the financial counselor, tap-tap-tapping on the calculator, "you will still have to pay $215 every month for the medications. Can you afford that?"

"For you," said the nurse, glancing at the chart, "it will be another year. We just don't get that many"—glance at the chart again—"O-type organs."

"African Americans reject kidney transplants more often. Their immune systems are just so strong," said the transplant nephrologist, his arms and fists crossing before him in a weak muscle pose, like he was paying Robert a compliment.

Each statement was like a sobering slap to the face. Perhaps since there simply weren't enough kidneys for everyone in need of one, it was the transplant team's practice to tailor their discouraging remarks to each patient who sat in that chair. But being privy only to what was said to us, the message we took away was "The kidney transplant system doesn't like Black people."

National data said it wasn't just our imagination or where we sat. I soon learned that though Blacks and Whites each made up a third of the kidney transplant waiting list at that time, Whites received every other donated kidney and Blacks received every fifth one, which meant that on average, Blacks waited nearly two years longer than Whites for a kidney transplant. As a primary care doctor at the time, not aware of the realities of nephrology, I didn't know that two years could mean never having to be on dialysis at all. That two years could be the difference between surviving in body and spirit. Or not.

There are a number of reasons why not everyone healthy enough to receive a kidney has an equal chance of getting one. A person must be in the care of a nephrologist—one who looks at *them* and considers transplant as a viable option and actually makes the request for a transplant center to evaluate them. According to several large research studies, only one or two out of every ten people with significant kidney disease are aware that they have it, and would, therefore, unlikely be in a nephrologist's care. There is no national repository of kidney function tests that generates automatic referrals for kidney transplant evaluation, and there is no one to ask a nephrologist why he or she hasn't referred a particular patient with advanced chronic kidney disease for evaluation.

Robert had been under the care of a nephrologist for years prior to reaching end-stage kidney disease but didn't hear about kidney transplant as an option until months after starting dialysis. Now in all fairness, few nephrologists would begin a conversation about kidney replacement options before the patient's kidney function—their estimated glomerular filtration rate or eGFR, how fast the kidneys filter the blood—had fallen to 25 milliliters per minute, and the patient cannot be placed on the kidney transplant waiting list until kidney function reaches 20. The medical world tends to simplify the explanation of eGFR to patients as "percent function," though 100 percent function of normal kidneys in a young person can be closer to 125 milliliters per minute, so a more accurate and easier-to-understand explanation would be that we start out with about half a cup, or 25 teaspoons, of blood flowing through our kidneys' filters every minute. Using this analogy, nephrologists usually start talking about kidney replacement options when it has dropped to 5 teaspoons a minute, and a patient can be placed on the waiting list once his kidney function has dropped to 4 teaspoons a minute.

So maybe Robert's kidney function dropped from an eGFR of 30 to starting dialysis (an eGFR between 5 and 10) in a year—unlikely, but it could have happened and would explain why his nephrologist hadn't mentioned anything. But Robert wasn't referred for transplant evaluation until he had been on dialysis for a full year. Since time on the waiting list is one of the most important determinants of who gets the next deceased donor kidney that comes in, when a person's name is added to that list can truly be the difference between life and death.

Years later, a transplant nephrologist said to me, "It's like rearranging deck chairs on the *Titantic*," suggesting that no matter how the rules for deciding who gets a kidney were changed, it wouldn't make much of a difference. There simply weren't enough kidneys for everyone who needed a transplant to get one. Somebody would be left out. And the ship was still going down.

It is true there aren't enough kidneys to go around—in the United States only about 17,000 kidneys are transplanted each year though more than 100,000 people are waiting for one—*But how about we unlock steerage to at least give the poor people a shot at a chair?* I thought. *Or, even better, a shot at getting into a life raft.*

In December 2014, the Kidney Allocation System implemented a new rule that turned the key in the steerage lock: the waiting list date would be backdated to when the patient started dialysis rather than at the time of transplant evaluation. This was an important step toward minimizing the effect of human error on access to kidney transplantation.

But it didn't go far enough.

It didn't go far enough because most people start dialysis when their eGFR is somewhere between 5 and 10. The difference between an eGFR of 20 and 10 alone can be several *years* for many patients. Years that could be spent accruing time on the kidney transplant waiting list. Sometimes so many years that a patient could potentially get a *preemptive* transplant—a transplant without ever having to go on dialysis.

No doubt how we estimate kidney function similarly delays how soon a patient gets on the waiting list, as with a patient I wouldn't meet until years later whom I call Book of Eli because he is Denzel Washington's doppelgänger as he appeared in the movie *The Book of Eli*, minus the dark glasses and shotgun. He even had similar mannerisms. The way he held his mouth. His laugh. How he talked. In short, he was a good-looking brother with a chronic kidney problem. Book of Eli was quite muscular when we met. But over the years, as his kidney disease worsened, his muscles had begun to waste away. When we met I needed a large cuff to take his blood pressure. Five years later, a regular-size cuff would do.

Like most of my patients with chronic kidney disease, when Book of Eli came in for a clinic appointment, he wanted to know how his kidney function lab test turned out very soon after "Hi how are you?"

I scrolled down his electronic medical record to find his creatinine and eGFR. Creatinine is a breakdown product of our muscles, is produced at a fairly constant rate every day, and passes freely through the kidneys' filters but, for the most part, is not pushed out or pulled back into the body by the kidneys' tubules. For this reason, we use it to estimate kidney function. Getting closer to *actual* kidney function requires both a blood test and collecting a patient's pee for twenty-four hours. Getting even closer to actual kidney function requires a research lab, an IV, repeated blood draws, and essentially all day—and this isn't even the *actual* kidney function.

In 1999, a bunch of researchers published a study of about 1,600 adults examined in order to come up with equations to estimate kidney function. Just plug in the patient's creatinine, age (because adults tend to lose muscle mass as we get older), and gender (because men tend to have more muscle mass than women), and *voila!*—an estimate of kidney function. Most laboratories can do this for us now. A rising creatinine level in the blood means the kidneys are not able to pee creatinine out as well as they used to, so the person's estimated kidney function is lower.

But wait—if the patient is Black, the study determined that you have to multiply by 1.2 to get a more accurate estimate. This finding was attributed to Blacks in the study having higher muscle mass than Whites and, therefore, higher amounts of creatinine in their bodies. Laboratories report the eGFR, and just below it, the eGFR if Black.

Of course one of the problems with generalizations is that they aren't always true. In medicine, in particular, they make us lazy and we often accept them without question—especially when they are in line with our underlying assumptions and beliefs. Like the belief that Black and African are inherently different from White and European at a DNA level, a belief that dates back to the days when American researchers were measuring Black-White differences in skull size to prove Black inferiority and justify slavery.

But I wonder how often health-care providers make the mental adjustment that the "race adjustment" is really a proxy for muscle mass rather than just focusing on the race of the person in front of them when they are assessing lab results. I wonder if the person in front of them were a White male bodybuilder how many would tell him the race-adjusted estimate of kidney function, or a skinny Black woman the non-race-adjusted estimate.

Then too I wonder how many health-care practitioners realize that equations derived from the original study of 1,600 people only included about 200 Blacks—and no American Samoans, no Hispanics, no Asians. These groups have very different body frames, but all are simply "not Black" in our equations. The implication, then, is that only Black people are different. This shortcut has the potential for a significant negative impact on Black patients who happen to not have a high muscle mass. Patients like Book of Eli. When the non-race-adjusted eGFR is 20 (when a person can be placed on the waiting list), the race-adjusted value is closer to 25. Just as the difference between eGFRs of 20 and 10 can be several years for many patients, so can the difference between 25 and 20. Years of accruing time on the kidney transplant waiting list when thirteen people on the waiting list die every *day* waiting for a kidney.

I explained this to Book of Eli and asked him to collect a twenty-four-hour urine sample and get a blood test so that we could get a more accurate measure of his kidney function. The twenty-four-hour test result was 20—the same as the "not Black" eGFR. His race-adjusted eGFR was 24. I referred him for transplant evaluation with the twenty-four-hour eGFR estimate. It was another two years before he started dialysis—with two years of time accrued on the waiting list. Two years closer to a kidney transplant. It probably would have been just a year if I waited for the race-adjusted estimate to dip to 20. Zero years if he wasn't under the care of a nephrologist. I was glad that Book of Eli was under my care, but especially glad that I could see beyond his race.

I knew none of this when Robert and I sat alone in that clinic exam room after the last transplant team member filed out. I just knew that the man I loved needed a kidney and it felt clear to me that the kidney transplant system couldn't or wouldn't move fast enough to get him one.

But I had one to spare, I thought. So it was in that moment that it first occurred to me to say, "We should try to see if I can give you a kidney."

"No, I don't want to do that," he said softly without pause and without looking at me. I was reminded of the last time I said something so startling that provoked a similar response.

"Let's make a baby," I had gushed then.

I was sitting on Robert's thighs, my knees leaning into the sides of his belly. He was sitting up in my bed, his back against the headboard, legs stretched out beneath me. We were in that sweet spot of a relationship where the love was so new that we smiled at each other in the miracle of requited love, but not so new that we hadn't worked through our first tiff.

Still, the words poured out of my face before I could stop them. It was my spontaneous proclamation of how big the love I felt for him was, of what was happening between us. How I wanted to tether myself to him in a way that was bigger than the two of us.

His smile gave way to brows lifted in surprise, then furrowed in worry. There were a few twitching beats at his right temple. He looked up and to the left, blinking, the way people sometimes do when they are trying not to cry. His green eyes glistened.

"I can't do that," he said, just above a whisper.

It was then that I saw it. I saw the uncertainty well up in his eyes. Uncertainty that did not allow for planning for a lifetime when dialysis gave him a three-times-a-week reminder that he might not get to grow old.

But what if he *did* get to grow old? What if he was one of the lucky ones who survived decades on dialysis? What if he *did* get a kidney transplant and his body didn't reject it, kill it, because that's what the body's immune system is supposed to do when something foreign comes in? He couldn't allow himself to hope for a transplant, but part of him needed to live like it would. But to plan for another life? He wouldn't allow himself to take it that far. That felt irresponsible.

I held his face in my hands and brought his forehead to

mine. He didn't need to say any more out loud. We exhaled deeply together as if there wasn't room in our lungs for our short breaths in. Everything would be all right. I would believe it and hope for it in a way that he couldn't. And there would be no more talk of babies until everything was.

In that clinic exam room, I saw the same twitch at Robert's right temple and the uncertainty in his eyes again. It was bad enough that he had brought me into his reality and captured my heart. He could not allow me to risk my life *too.* If something went wrong, he wouldn't be able to bear it. That would be too much.

And then there was the possibility of disappointment. That I might change my mind. I knew a few before me had said the same words to him—*Let's see if I can give you a kidney*—but each had fallen by the wayside for whatever reason. Some offered that there was a medical condition that would make it too risky. Some just tiptoed backward into the shadows. Robert never asked why. He didn't want to see the truth that he suspected in their eyes—that they wanted to give him a kidney *in theory,* but not when reality set in.

Robert didn't ask anyone to consider donating. He didn't want to be disappointed in people.

"It's not fair to be disappointed in someone for not being able or willing to give you an organ. It's just not fair," he offered years later. "And what if I couldn't do all the things I needed to do? Then we would both be fucked up." He didn't want the responsibility of a kidney from someone still walking around.

But in that moment, I didn't need him to say it. I made up my mind to not say another word about it to him until I knew if I was a match.

As I went through the process of finding out if I could donate my kidney—blood tests to confirm that we were a compatible blood type (type O) and that I had no signs of diabetes or high blood pressure; a physical exam to confirm I had no signs of disease (minus the rectal exam, which I refused to allow—"I am not middle-aged and I don't have a prostate. Your finger does not need to go there," I said to the petite nephrology fellow, and she backed down without further fuss); a twenty-four-hour urine collection to confirm that there was no blood or protein that hinted of kidney disease; and finally a CT scan to confirm that I actually did have a kidney to spare (and wasn't the 1 in 750 born with just one kidney, or the 1 in 400 born with two fused into one horseshoe-shaped kidney)—I told family, friends, and colleagues of my plans. Most were supportive, impressed even, at my willingness to do such a thing.

"It's not an extra pair of shoes, Vanessa," said one awed colleague in response to my attempt to shrug off praise with a nonchalant "He needs one and I have two."

But not everyone was so supportive.

"I would do it for my partner in a heartbeat, but we've been together for more than a decade," said another colleague. Translation: *Girl, you don't even have an engagement ring. Are you crazy?*

"Can't somebody else do it?" asked my devout Christian sister.

*That ain't Jesus*, I thought, because her brand of devout was *beyond* just plain devout. It was the kind of devout that stood on a street corner with a bullhorn preaching the gospel and traveled to Africa rapping it. I was surprised that beyond-devout's love and charity for all God's children seemed to end with immediate family in this case.

Truthfully, it didn't matter what any of them said. I was going to do it, because my heart was invested. I wanted, *needed,* Robert on the planet for as long as possible. I knew that my fresh, living kidney, as opposed to one from someone who had died hours earlier, would give him the best chance that "as long as possible" would indeed be long. I believed at my core that giving Robert a kidney was the right thing to do and that everything would be all right.

I was sitting in a pediatrician's exam room when I got the sign that my belief was true. Avery was four years old and at his well-child checkup. He sat on the floor moving green, yellow, red, and blue wooden balls around a maze of yellow wires, his hands and mouth in constant motion as usual. I picked up an old magazine from a basket on the floor. A *Woman's Day* or *Family Circle* or the like, just something to flip through to help pass the time. But . . . then . . . about four flips into it . . . there it was: a story about the fiftieth transplant anniversary for The Other Robert Phillips. There was my sign.

It never occurred to me that anything would go wrong. Of course the surgery would go smoothly and Robert would live free of dialysis for the rest of his life. Never occurred to me that I would be the one in ten thousand who died trying to donate a kidney, leaving my four-year-old motherless. I'd be three times more likely to be struck by lightning over my entire lifetime, so of course I'd be back to Avery in just a few days, I believed. It didn't bother me that I'd lose half my kidney function until my remaining kidney had time to compensate for its missing mate. I didn't even know that about two weeks later my right kidney would plump up to provide about 70 percent of my total original kidney function and up to 85 percent over the long run. All I needed to know was that I could live just fine with my one kidney as long as I continued to take good care of myself by not getting so fat that I developed diabetes or high blood pressure or by not starting to smoke cigarettes or crack or shoot up heroin.

I wasn't aware that the kidneys were involved in fertility. I knew other women had babies after donating. After the transplant surgery, of course Robert and I would make that baby I hoped for.

Incredibly naive, I know, but also not surprising that I felt so sure of the outcome, as I tended to be a glass-half-full person in my day to day. But somehow I've always just expected to get what I really want in life—from college and medical school acceptances to residency program and after-training job offers even though I was far from a star medical student or resident. I didn't always get what I wanted, at least not at first, and occasionally not at all. And each time I didn't get what I wanted, I wouldn't just be disappointed, I would be dismayed, incredulous even, like an early *American Idol* reject.

I remember the day I found out that I could be Robert's donor. After multiple phone calls trying to find somebody who could tell me, the news was given to me in very basic terms—there was no clotting when my blood was mixed with Robert's. We were, therefore, compatible. Of course we already felt compatible at the relationship level, but it was good to hear we were compatible at the DNA level too. It was years later that I learned we were a match at only one of six genetic markers considered at the time. Maybe I wouldn't have been quite so cocky had I known this detail. In that moment of knowing only that the answer to the simple "compatible yes/no" test was in our favor, I burst into happy tears.

Back in pre-op, I smiled to myself as I thought about how far we had come since dinner at Melanie's just nine months earlier. My skin felt tingly with my nervousness, yet I felt at peace. The thick blue curtain separating me from the patient to my left slid abruptly down its metal track as my nurse appeared with a syringe in her hand.

"OK, almost time to go. Here's 10 milligrams of Valium," she said, smiling, as she pushed the clear liquid into the IV at the crook of my left elbow.

I smiled back at her and shivered a little, more from her words than the cool surge in my vein. *It's almost time to go.*

Moments later, three young surgeons swarmed in and unlocked the wheels of the bed I lay in. Click. Click. Click. Click. They started to roll me toward the operating room.

"Are you feeling sleepy from the Valium yet?" asked one, his clean-shaven beige face hovering above me.

"No, not at all," I said, feeling wide awake and wondering if they would have me count backward from twenty with the mask over my mouth and nose as I inhaled the vapors of anesthesia, like I'd seen them do with patients when I was in medical school.

# 5

## PEANUT IN A BOX OF WORMS

Operating Room 17 was like a refrigerator, but I was unaware. I was unconscious and paralyzed.

A plastic mask was taken away from my face and a hand tilted my head so that my jaw jutted forward. Two fingers, in an exaggerated snap, pried my mouth open to make room for the sickle-shaped blade of the laryngoscope. Once it was in place with its tip at the back of my tongue, the left hand lifted the handle so that my vocal cords were exposed. A breathing tube slid between them into my windpipe and the metal rod that gave the breathing tube shape was removed. With the bell of a stethoscope against my rib cage on the right, then left, the anesthesiologist listened as air moved through each of my lungs. She was satisfied that the breathing tube was correctly positioned. She connected the end of the tube to the machine that would do my breathing for me and secured the tube with white tape to my face. The machine displayed my blood pressure. My heart rate. The settings from turned dials that continually pumped drugs through the IV to keep me unconscious and immobile while my left kidney was being removed.

The kidneys are like the Rodney Dangerfield of vital organs—they get no popular respect. I blame Hollywood and the music industry. With the exception of that scene in *Steel Magnolias* where Shelby told everyone in the beauty shop that her mama was giving her a kidney so she could stop looking like she was driving nails into her arm for dialysis, the kidneys have been pretty much glossed over. We have countless love songs about hearts aching and breaking, brains that can't stop obsessing, and lungs that can't breathe again—but not one about the kidneys. Nobody was singing "If I only had a kidney . . ." in *The Wizard of Oz*.

But it wasn't always this way.

Early translations of the Bible valued the kidneys. The *reins* as they were called then, from the Latin *rēnēs,* were central to the soul, the stimulus for moral yearning and righteous action. According to the Masoretic text in the Holy Scriptures: *"I will bless the Lord, who hath given me counsel; Yea in the night seasons my reins instruct me."* It was through the kidneys that God examined the moral worth of Man: *"Examine me, O Lord, and try me; Test my reins and my heart."* And they were the target of punishment for immoral behavior: *"He cleaveth my reins asunder, and doth not spare. . . ."*

And Hippocrates, the father of medicine, believed the kidneys' secretion of urine held the most important clues to what was making a person sick. And just like people think "physician" when they see a white coat and stethoscope today, it was the *matula,* with its clear glass bottle rounded at the bottom, like a bladder full of urine, held up to the light that symbolized physician well into the Middle Ages. Pee was the shit.

But then things changed.

The kidney and its secretion fell from the virtuous pedestal they enjoyed for thousands of years to one of ridicule and caricature—all because too many physicians tried to do too much with their patients' pee. Many stopped bothering to examine the patients themselves because their urine alone was all that mattered to them. Translations of medical texts from Latin allowed those with no formal medical training to get in on the action, with lay healers diagnosing disease solely by the color of the urine. Soon they began telling fortunes and predicting the future—by urine. Physicians joined in. Even witch hunters started using urine to identify their targets. Eventually it became so ridiculous that a medical rebellion against all uses of urine for diagnosis ignited. In the early 1600s, *The Pisse Prophet* was published. It called out all those who pretended knowledge of disease by urine as engagers in "piss-pot science." This reputation stuck. More than a century later, a caricature depicted a dozen "quack heads," one of whom had his finger in a matula—literally poking fun at the pseudoscience of disease diagnosis by urine. The kidneys came to be viewed as mere producers of waste. Modern translations of the Bible replaced the reins with the "mind" and "soul."

It wasn't until kidney transplantation became a reality

that the kidneys began to regain even a portion of their past glory. But even though the kidney was the first solid organ transplanted, the concept of transplantation was nothing new. According to folklore, transplantation was attempted as far back as the second century in ancient Rome, when twin brothers and physicians Cosmas and Damian transplanted the black leg of a dead Ethiopian man onto a patient's white body. Supposedly the patient, whose own leg may have been destroyed by ulcers, did well. The brothers, on the other hand, were beheaded, but this may have had more to do with martyring themselves for their Christian faith than the transplant.

Nevertheless, kidney transplantation didn't really get started until 1900, after overcoming several technical hurdles. French surgeon Alexis Carrel is credited with figuring out how to connect the blood vessels that would allow blood to enter and leave the transplanted kidney through a series of transplant experiments in dogs in 1902. Before him most surgeons used glass tubes to connect blood vessels. In 1905, N. Floresco, a physiologist in Bucharest, was first to suggest connecting the donor ureter to the recipient bladder. However, for many years, surgeons feared the stitches between the two would come undone, so they connected the ureter to a hole they made in the skin. Early experimenters transplanted kidneys into arms and thighs before landing on the pelvis. Since kidneys in these early experiments actually produced urine for only about an hour, I imagine no one had to consider the awkwardness of having to pee out of one's armpit.

They didn't realize that the rapid failure of kidneys was

because of rejection—the recipient's immune system behaving as it should, destroying the invader, not realizing the invader had come in peace with the intention of saving its life. This was the fate of the first human-to-human kidney transplant, performed by Dr. Yu Yu Voronoy in the Soviet Union in 1933. The recipient died two days later. The first successful human kidney transplant was performed by Dr. Joseph Murray in Boston in 1954 when Richard Herrick received his identical twin brother, Ronald's, kidney. Here rejection wasn't an issue because it was like moving a part to another location in the same body. It was this avoidance of the immune system altogether that allowed successful kidney transplantation to depend upon surgical technique only, a technique that has changed very little to date.

It would be another several years, with the discovery of methods of defeating the immune system, before any extended survival could be achieved in the greater population. The first was radiation, but because it killed the entire immune system, the body was left with no natural defenses and as a result would succumb to horrible infections. Next was azathioprine, a derivative of mustard gas. Initially it was considered to be an antibiotic, though a very weak one because it actually weakened the immune system—but this was a feature that could be exploited for use in transplantation. The addition of prednisone a few years later allowed lower doses of azathioprine, which meant lower infection risk, but still kidneys lasted just shy of two years. It wasn't until cyclosporine was discovered in 1976 that the balance between preventing rejection and avoiding infection greatly improved. That plus simultaneous efforts to figure out how to better match donors to recipients and the discovery of newer, better drugs over time has created the current reality in which more than two-thirds of all kidney transplants are still working after five years, while little more than a third of dialysis patients are still alive in that same time span. Some kidney transplants last twenty, thirty, forty, even fifty years.

I hoped Robert's new kidney would be among those still working after fifty years.

My pale blue gown was whisked away and I was naked for all to see. But to them, I was just another body. All that mattered was how much fat there was to cut through, how much pee was coming out of me, and that I woke up in the same condition I went to sleep in—minus one kidney.

Gloved fingers spread my labia apart. Swab, swab, swab, the area was cleaned. The narrow, floppy yellow catheter was lifted out of clear lubricant and its end was guided to the opening of my urethra and pushed in deeply enough for its surrounding balloon to be inflated with saline to keep the catheter from slipping out of my bladder.

The transplant surgeon, Matt Frank, walked in as if the plunge of the syringe filling the balloon was his cue. He wore metal-framed glasses, a surgical mask, and a periwinkle bouffant cap like mine that covered a mane of thick silver hair. His presence added an intensity to the room. His surgical assistant, the fellow who had completed his general surgery residency and was now learning the specialty of transplant surgery, suddenly stood taller and moved into action. He, Frank, and the nurse worked quickly, methodically, to position me. The nurse attached a brace perpendicular to the right side of the table and lined it with thick blue foam. My right arm was laid in its cradle. I was rolled onto my right side and more foam was placed along my backside. The nurse attached another brace to the bed and lined it with more foam for my left arm to lie on, suspended above the right. I looked like the mummy of old movies tipped over. Wide white cloth tape wrapped around, securing it all in place. Foam between my knees and ankles. More white tape. I would be like this for two hours. All the careful padding ensured that no nerves would be pinched, so that when the anesthesia wore off, I would be able to move all my parts.

A swab soaked in brown Betadine landed on my belly. The fellow's hand swept the swab in ever-widening concentric circles until I was painted a deep yellow from just beneath my breasts to just below my C-section scar, now traced in purple from the surgical marker. A new soaked swab painted again. A third for good measure, to ensure that all the bacteria naturally living on my belly skin was dead. The fellow wondered why Frank still used Betadine, while the other attending surgeons he scrubbed in with used a clear alcohol-based solution.

As if reading his mind, Frank said, "I like Betadine because it doesn't wash off our markings," though part of him realized this made him seem old school.

I was left to air-dry. Then a thin plastic film impregnated with iodine—just in case the Betadine didn't get all the bacteria—was stretched over my belly like yellow plastic wrap over a leftover ham.

Frank and the fellow stepped back to don sterile gowns and gloves. The scrub technician was in the corner, already cloaked in sterile garb. He arranged all the scissors, clips, clamps, needles, and gauze sponges rolled like cigarettes on the table before him. Counting. Counting again.

The nurse held an open pack for the fellow to retrieve its sterile thick blue paper contents. The paper was spread all over me so that only about a square foot of my belly skin was exposed under the transparent yellow plastic. The table was tilted so that my left hip was higher than my right.

A laparoscopic left nephrectomy was under way.

Four horizontal cuts were made on my belly, each about an inch across. Three in a line about three inches apart like nipples on a sow's belly. The fourth about three inches left of the middle cut.

I was lucky. Had I donated a kidney a decade earlier before laparoscopic nephrectomy had been developed, I would have had an open nephrectomy—pretty much sawed in half, waking up to an incision a foot or so long and possibly one lower rib lighter. I could have expected to spend up to a week in the hospital, be in a lot more pain, and need an extra month to get back to work. With the laparoscopic technique, I might be in the hospital as few as two days and back to work in a couple of weeks. I could even dare to dream that I'd wear a bikini again.

A hollowed-out needle connected to plastic tubing punched through the first incision and my belly was filled with air. A trocar, something that looked like one of those jumbo pencils for preschoolers, but hollowed out and with what looked like a spinning top, replaced the needle. A camera was passed through it. It magnified my insides almost fifteen times and displayed them on a twenty-one-inch monitor.

Three more trocars were pushed through. One through

each cut. Through each trocar the instruments were passed. Little grabbers and scissors at the ends of long metal rods. The surgeons' wait for each instrument was on the order of milliseconds as the scrub tech anticipated their every move. They watched the monitor as they moved the instruments like puppeteers manipulating my organs and the stuff between them.

It was all shades of pink with some areas of iridescence reflecting the camera's light. All of it looked like one vague indistinguishable mess to the average person. But Frank and his assistant worked as if everything was labeled like in the medical school anatomy books. They worked steadily, identifying my structures almost in whispers as they singed through planes of tissue. Spleen and large bowel were moved aside to bring my left kidney into view. They traced its ureter still working to carry newly generated pee to my bladder. They slung a plastic loop around it so it could be quickly found later. Back up to the kidney's vein and artery. Plastic loops were slung around them too after all the extra was cleared, including two veins sprouting off the kidney's vein. Robert wouldn't need my gonadal and lumbar veins, so they were clamped and cut. All fluff removed.

Other than the *beep*, *beep*, *beep*, sounding throughout each singe, the room was quiet. Words were spoken only as needed. Including names. Everyone simply knew when he or she was being spoken to. All fluff removed.

"Look the camera back up to the spleen," Frank instructed the fellow.

With his left hand, the fellow directed the camera as told.

"Take that down there. . . . No, like this."

The fellow corrected himself as he etched each lesson into his brain.

"How much urine?" Frank asked, his voice a little louder.

"Ten cc's in the last fifteen minutes," reported the nurse without hesitation, as if she were anticipating the question.

"Is the Mannitol in?"

"Another twelve point five milligrams going in now," from the anesthesiologist.

"Follow that with another twenty of Lasix."

"Yes. Twenty milligrams of Lasix going in now."

"Let them know time to kidney out is fifteen minutes."

The nurse left the room to carry out the order.

More clearing, isolating the kidney. Another cut over the one made to pull Avery out five years earlier. My belly deflated. Tiny bleeding capillaries peeking out from my belly fat singed along the way. Then a stillness as they waited for Robert's surgeon, Shun Kobayashi, to enter the room.

I remember meeting Kobayashi from the transplant coordinator's office the week before surgery. He shook Robert's hand and then mine. I noticed the short, stubby fingers that didn't reach past my palm. They matched his short, stubby frame, but not the image I had in my mind of what a surgeon's fingers would be like—long and almost elegant like a pianist's. He seemed pleasant enough, so Robert liked him, because it was a good thing to like one's surgeon, he thought. But we both thought how odd he looked in his green surgical scrubs topped with a tweed jacket instead of a white lab coat. He looked like he was trying to present an unnecessary formality but that he really didn't want to fully commit to it by getting fully dressed. It made me not trust him, but Robert dismissed it as just a manifestation of the typical nerdiness all doctors shared to some degree or another.

Now Kobayashi was donned in sterile gown, sterile gloves, surgical cap, mask, and a headpiece fitted with little spotlights and magnifying lenses. Close to his chest he held a flat-bottomed bowl filled with a clear icy slush in his hands like it was a blanket waiting to catch a newborn baby. He stood opposite Frank and the fellow.

As he walked in, a new sense of urgency overcame the room. There were no words. Just movement. Snip, snip, snip, in rapid succession, and my kidney was freed. Now the passing seconds were counting the time my kidney had no blood flow. With each passing second, more bits of my kidney were at risk of suffocating. Dying. No longer working. Metal rods out, out, out. One last look at my kidney with the camera. Then a sudden thrust of Frank's gloved hand, then forearm, then almost elbow, through my C-section, soon-to-be nephrectomy, scar as my belly deflated, taking the camera's ability to project the visual field with it. Frank's eyes were fixed, concentrated but not seeing straight ahead. Reaching. Finding. Grasping. Pulling. He was determined to grasp my slippery kidney before it had a chance to drop down into my right side, at which point he knew finding it would be like finding a peanut in a box of worms—a four-inch kidney in twenty feet of small intestines. But he had it. He exhaled slightly, relieved.

Out came my kidney, fully contained in the palm of his hand. It was smooth, dark pink, and shiny with bits of extraneous tissue hanging from its artery, vein, and ureter.

Toss!

Into the icy bath in Kobayashi's arms it was submerged. It was Robert's kidney now. And with a near-military precise left-face turn, Kobayashi was on his way out of the room back to Operating Room 18 next door, where with his fellow less than two minutes earlier he had been clearing a space in Robert's pelvis for my old kidney and his new kidney to lie.

With the excitement over, attention turned back to me like stunned restaurant patrons returning to their meals after a fighting couple stormed out. The fellow methodically sewed my C-section/nephrectomy incision closed, then filled my belly with air so that the camera could once more show them that all the clips were holding and there was no more oozing in need of cautery. Camera out. Air out. Final stitches to close the tiny incisions. And after all this I'd lost less than 4 tablespoons of blood.

*Vanessa. Vanessa.* Suddenly I heard a woman's voice piercing into my consciousness. Pleasant. Singsongy. It must belong to the person who was nudging me.

"You're all done," she said.

I opened my eyes and looked at her, surprised to see this new face when it seemed I had just blinked at the elevator a moment ago.

"You're in the recovery room," she said brightly.

It was at that moment—and not until that moment—that I realized the bigness of what I had done.

I smiled back at the nurse, happy to be awake.

# PART II

## BUT WILL SHE STAY

# 6

# COMPLICATIONS

I looked to my right and saw Robert in the surgical recovery bay next to mine. He was smiling at me, finally believing a transplant was happening for him. I smiled back.

I thought of the others who had considered this same path and had maybe even spoken their intentions out loud, but then retreated long before reaching that point of no return. I wondered why retreating hadn't even occurred to me. Was I that blinded by love? Was I that naive? Was I trying to earn my own forgiveness for past mistakes?

Yes. Definitely. Maybe.

But in that moment I felt thankful that retreating hadn't occurred to me, that I had been so bold, regardless of the reasons why.

My eyes left his to look for the bag collecting urine from his Foley catheter. I found it hanging by a plastic hook at the foot of the bed. Pale yellow urine—and a lot of it—was there. The kidney was working beautifully. I smiled bigger, with teeth showing, and brought my eyes back to his. He saw the happy in my face and matched my smile. We reached for each other, though we were too far apart to touch fingertips. But it didn't matter, because in a way we were already touching. We were tethered, connected through our new state of sharing two healthy kidneys between us, replacing his old tether to the dialysis machine. We settled for a fingertip wave.

"I feel like the fatigue has just lifted off me," Robert said, amazed. He didn't realize how tired he had been. He thought it would be several days before he'd notice a difference.

I smiled bigger, even though I felt like all of his fatigue had been dropped onto me. This was new to me. The most my healthy body had endured before this day was a C-section to bring Avery into the world.

Robert, on the other hand, had endured kidney failure and all that came with it. Having a dialysis catheter inserted. The catheter becoming infected. Sepsis, the infection spreading into his bloodstream. The catheter being removed. A new catheter put in. A surgery to create his fistula. Another surgery to revise his fistula when it wasn't working properly. Nausea. Vomiting. Leg, back, jaw, hand, everywhere cramping. Dialysis Monday. Dialysis Wednesday. Dialysis Friday. For almost six years.

But now he had a healthy kidney.

Some hours after surgery, after we had each been moved out of recovery to our separate rooms in the kidney transplant unit, I was happily pushing the patient-controlled analgesia button as often as I was told I could. Then, just before midnight, Robert walked into my room. He was with a nursing assistant and using a walker and wincing a little, but he was *up*, *out of bed*, and *walking* the evening after having an abdominal surgery as I lay in bed pushing the PCA button every few minutes. *Wuss*, I thought of myself.

It was like watching a woman in labor refusing all pain medication while I was calling for the anesthesiologist. In truth, I was that woman calling for an anesthesiologist when Avery was born. Avery was in distress—butt first and pooping—so I would have had to have anesthesia to be rushed off to C-section anyway, but after a couple hours of that blinding white pain every few minutes, I was asking for the epidural long before anybody knew of Avery's distress. I started out wanting to feel the experience of labor in its entirety—dilating cervix, pushing, all of it. I felt my cervix open from three centimeters to four. *Good enough,* I thought, and exhaled calmly, "Anesthesiologist, please," when the white before my eyes receded into color again.

"Hey," Robert said, smiling brightly in the doorway.

He padded slowly to me and leaned over to plant a kiss on my lips. He grimaced a bit as he straightened, but still smiled because the pain was because he just got a kidney. Things were good. He could feel how it would be having a normal kidney working all the time rather than the little bit of kidney in the form of a dialysis machine a few hours each week.

His eyes weren't red anymore. His skin no longer had that washed-out look. He had already been happily guzzling ice-cold water from the pink plastic hospital pitcher to his heart's content.

"They told me to drink," he said, grinning.

I grinned back. I was so happy for him. Gone were the days of having to limit how much liquid he took in between dialysis treatments. Now he had a kidney that could pee out any fluid his body didn't need. He focused on learning about all the new antirejection medications that would keep his new kidney healthy. He was eager to get on with making up for the six years he lost to dialysis. The six years he had to curb his ambition and planning for the future because he didn't know what the next hours, days, weeks, months, years would bring.

But on the second day, things changed.

The urine flow into the collection bag slowed to a trickle, then stopped altogether. Every few hours the nursing assistant would weigh him, and each time his weight would be two and four pounds higher than the last time. Something was wrong.

He was taken to the radiology department for an ultrasound that afternoon. It showed hydronephrosis—a significant amount of urine had backed up so that the kidney's pelvis, the central part of the kidney that normally funnels urine to the ureter (the tube that funnels urine to the bladder), looked like it was stretched to more than twice its normal size. Because the capsule enclosing the kidney will stretch only so much, the ballooned pelvis would press against the urine-making parts of the kidney. Pressed long enough, these parts would wither and eventually die.

Not long after Robert was returned to his room, a team of rounding doctors filed into Robert's room, with more spilling into the hallway. By the lengths of white coats and amount of gray hair, it was clear there was a pecking order. The long, knee-skimming white coats and white hair of attending physicians in front and short coats just clearing the waists of the blond and brunette medical students bringing up the rear, straining to hear what was being said.

"There is a blockage," said the transplant nephrologist leading the pack, his face concerned, his demeanor confident. "It's the reason why you're not producing as much urine as we'd like. We'd like to go in and see what's going on. The urologist is here." He gestured to the long-coated, salt-and-pepper-haired man to his left. "He's taken a look at your ultrasound and the blockage is somewhere in the connection to your kidney."

This was not supposed to happen. This was a surgical complication.

About a half hour before Frank walked into Operating Room 17 to find me, Kobayashi had entered Operating Room 18 to find Robert lying on his back where the kidneys he was born with, his native kidneys, lay shrunken like grapes withered on the vine, which is where they would stay. Only native kidneys causing problems are removed. Problems like an infection that won't go away. Or like the problems common to people who have inherited polycystic kidney disease, whose kidneys are nearly completely taken over by large cysts. Cysts that often bleed and cause pain and make the kidneys huge, like footballs, when a normal kidney is the size of a fist.

Robert was unconscious and paralyzed. Tubes already inserted. One down his throat for breathing. A smaller one in his neck for delivering the first antirejection medicine into his bloodstream. Another threaded up his penis to collect all the urine that was soon to come. Kobayashi's assistant, the surgery fellow, had already shaved and cleaned Robert's belly. He was ready for sterile drapes to be laid across.

Kobayashi expected that his assistant and he would make their incision about an inch above his right hip bone and cut diagonally down parallel to the thigh crease, then methodically cut through fat and muscle and tease apart connecting tissue until they had clearly exposed the parts of the pelvis they were looking for—a length of large artery and vein near the bladder. Parts ready for transplant.

Minutes later Kobayashi would return with the new kidney resting soft and pale pink in a bowl of cold slush and head to the back table, where he and his assistant would get the kidney ready to transplant. This wouldn't take long—kidneys from living donors don't require as much work to isolate their artery and vein, tie off extra vessels, or strip away a lot of fat like deceased donor kidneys require. All those extra parts make it harder for the living donor's surgeon to pull the kidney through a small bikini-line incision.

Not trusting the kidney in the hands of the assistant, Kobayashi would carry it himself to Robert's awaiting pelvis. He would lay it in the wound to judge its best position, given the length of transplant artery, vein, and ureter he had to work with. Once decided, his assistant and he would peer through magnified lenses mounted to their headpieces fitted with spotlights and sew the finest of stitches to connect vein to vein, artery to artery. Hands steady. No room for error. When the clamps were unclamped, he expected the kidney would be already "pinking up" with blood flow restored. It would be firmer too. They would watch urine dribbling out of the ureter. *This is a good kidney,* they would think.

Next they would attach the donor ureter. At this hospital, they almost always attached donor ureter to recipient bladder. Today would be no different.

"OK, ready to distend the bladder," Kobayashi would instruct, and the anesthesiologist would turn the clamp on the bladder catheter to "fill" position, and an antibiotic-laced fluid would flow into Robert's bladder. Though it hadn't had to fill with urine in the more than five years that he'd been on dialysis, with just under two cups of fluid, the bladder would appear like a balloon being blown up into view. Next he would cut a hole in the bladder, sew the transplant ureter into it, and wrap the two with a strand of muscle. This would both stabilize this new connection and mimic nature.

"Ready to unclamp," he would say, and the anesthesiologist would turn the catheter clamp a quarter turn. Urine would flow.

However, instead of a smooth dissection down to vessels and bladder, Kobayashi found layers of scar tissue and mesh stuck together in the right side of Robert's pelvis—perhaps not surprising given that he had two hernia surgeries on the right side. He found it almost impossible to get through. There was the left side, where they often operated for people who had prior surgeries or blood vessel blockage on the right, but that would mean starting a new dissection and the blood vessels on that side would be a little harder to work with—so connecting my ureter to Robert's ureter seemed the best option. Kobayashi rarely did this in his practice, but other transplant centers often did. He thought about putting in a stent, a wire mesh that would protect the new connection and hold it open. He tried, but the angle of the connection was too sharp. It was almost impossible to get the stent past it, so he decided not to persist. After all, urine didn't need a straight path to flow, so he thought it would be fine.

But on that second day after surgery, that connection was kinked like a garden hose that no water could flow past. The kink had caused urine to back up in the kidney. The kink had caused urine to leak into Robert's pelvis. The kink was the reason why Robert had stopped peeing and was becoming more and more swollen.

On the transplant nephrologist's cue, the urologist took over to explain what they wanted to do. They hoped to straighten out the connection between my ureter and his ureter without having to reopen his surgical wound. Robert listened intently, welcoming the notion of the problem being fixed.

"Let's do it," he said. And shortly thereafter, he was taken to the operating room.

Robert woke to *two* tubes coming out of his penis. A stitch attached a new narrow plastic catheter to the slighter larger one he went into surgery with. While he slept under general anesthesia, the urologists had gone through his penis with a rigid cystoscope, a hollowed-out metal rod slightly larger than a regular number-two pencil and configured with a light and camera that allowed them to pass surgical instruments through it and to see inside his bladder and ureter. What they found was a ureter in a tight Z formation, urine leaking at the second turn where my ureter was stitched to his. They were able to thread the new catheter around these sharp turns. All that Robert could see was the end of the new catheter with thin black lines around it. He was told that if he saw four of these lines, he was to push the catheter back in so the whole thing wouldn't fall out. This strange jerry-rigged contraption didn't seem like much of a fix, but Robert tried to be hopeful that the problem was indeed fixed since at least urine was coming out again.

However, a day later, the afternoon of the fourth day after transplant surgery, the swelling began to worsen. His hope waned. In hope's place, suspicion took hold. Was there something they weren't telling him? Why *couldn't* he see his chart?

With an IV pole supporting his urine bag connected to his penis tubing, he walked to my room, where I lay in bed twice as long as most donors. Though I had been walking around too, my bowels had yet to wake up after the anesthesia—and you aren't allowed to leave the hospital until you can pass gas.

Robert stood just inside the door trying hard to smile, trying to hold true to his stoic linebacker type of man reputation, but I could see the hurt in his eyes. As he stepped into the room, his eyes shifted to his mother, Ginger, who had come to sit with me after visiting with him.

Ginger sat in the room's recliner, her back straight and her legs crossed at the ankles like a proper lady. She was a petite woman with fair skin and wavy short hair. She didn't look like she could have given birth to two sons who would grow up to be over six feet tall and 220-plus pounds, particularly once you learned that she had rheumatic heart disease, which caused her to need a heart valve replaced when she was twelve years old and another when she was eighteen.

It seemed to me that their illnesses gave them an extra special mother-son bond, but the bond in fact was there long before Robert was ever tethered to the dialysis machine. After graduating from Morehouse College, Robert had planned to spend his summer relaxing and hoping his waiting list for Stanford Law School would turn into an acceptance to enter in the fall. Instead he spent the summer at Ginger's hospital bedside. At age fifty-eight, she needed surgery to replace the two previously replaced heart valves, and almost died in the process. Her heart stopped twice within a few days after surgery, and she didn't wake up for nearly a month.

When Ginger woke up, she was upset to learn how long she had been in the hospital. And she woke to collapsing lungs that had been repeatedly punctured by ribs that were broken from CPR. She needed another surgery to make them heal properly.

She went home with three wounds that needed to be cleaned and bandaged daily. One on her breastbone, which needed to be sawed in half for the heart surgery. Another over her windpipe, a remnant of the tracheostomy she needed to make room for the plastic tube that connected her to the breathing machine for such a long time. And a third on her back where they needed to go in to repair her ribs.

A home-visiting nurse would come out three times a week for those first two weeks to change the wound dressings. Robert changed the dressings and everything else between the nurse's visits and every day after the nurse's visits stopped and through that fall and winter. He didn't trust her care to anyone else, not even his father. Being on Stanford's Law School waiting list made it easier. There was nothing concrete to turn down. He didn't really want to be a lawyer anyway. It was just a path to the career in politics he secretly wanted.

He prepared her breakfasts and lunches. He cleaned and bandaged the wounds for the six weeks it took them all to heal. The tracheostomy scar was the hardest. It would be caked with green pus that needed to be cleared away with a swab dipped in saline. He bathed her. Seeing her naked was no big deal. He, his father, and his brother were used to seeing her walk around the house naked. "This is my house," she would say, and proceeded to walk around it any way she pleased, proudly displaying swinging breasts, pubic hair, and the scar from her childhood open-heart surgery that crossed her lower ribs, usually on her way to retrieve underwear from the dryer.

The bedpan did take some getting used to, but he emptied it until she was strong enough to use the bedside commode. And then he emptied that. As she got stronger, he took her to doctors' appointments and walked with her morning and evening.

It was during this time that their conversations went deeper than those between most mothers and their sons. She talked to him about things that surprised me, like about the many, *many* men who had proposed marriage to her before his father.

When she saw Robert appear in the doorway of my hospital room, she delicately put down the newspaper she held.

"You told her I broke down?" he asked her.

She nodded. Her nod was like a switch that opened the floodgates to his emotions. Tears rolled down his face.

I climbed out of bed as fast as my bloated belly would allow and went to him. I wrapped my arms around his neck and held him to me, not caring that the back of my gown had fallen open, exposing at least three-fourths of my ass to his mother. His forehead rested on my shoulder as his own shook with his sobs.

"I'm scared I wasted your kidney," he admitted. His worst fears were coming true. He was right not to hope, wrong to believe a transplant would happen for him, he was saying to himself. It would have been easier to bear had it been a deceased donor kidney. You don't have to look a deceased donor in the eye when things go wrong.

"It's not your fault," I spoke softly in his ear.

He knew it wasn't his fault, but that didn't matter. He believed his body had failed him. Again. Worst of all, he believed he had failed me too.

"It's gonna be all right," I followed with full conviction, still believing with my entire self that everything *would* be all right. I had to believe. He had my kidney and my heart.

# 7

## ANOTHER DIALYSIS MACHINE

It was the morning of my third day out of the hospital, and I was on my way back to visit Robert. I needed naps and was moving slowly, but I knew I would feel back to my normal self in about a month. And I knew that this experience—me the doctor as patient—would be good for me, the doctor caring for patients—particularly the experience of spending nights in the hospital. Lying in a bed while the faces of strangers hovered above me, asked questions like "Have you passed gas yet?" while a different stranger's hand lifted my gown without warning and pressed on my recently cut belly. The daily 4 a.m. checks of my blood pressure, pulse, respiration, and weight. The constant sharp, stabbing pain in my right shoulder, a common side effect of laparoscopic abdominal surgery. All of it gave me the ability to explain and empathize from the perspective of a lived experience, not just as a doctor who read about it in a textbook or watched somebody else go through it.

As exhausted and slow moving as I was, I knew things were much, much worse for Robert. It was his eighth day in the hospital. He hadn't wanted to talk much the night before. The urologist's attempt to make the Z-shaped ureter work had failed. His IVs had clotted off the previous day, and because he was still so swollen, the nurses couldn't find a vein to insert a new one. So he was hurting but no longer had an IV for pain medicine, and the pain pills only made him more constipated.

A different transplant surgeon had joined the team that day, and she would take him to surgery the next morning. Again.

I arrived at the hospital that day to find his room missing a bed. I went to the nurses' station to ask when they expected him back. He was out of surgery and back to recovery, but they were not ready for family to come in.

After a while, Sara Cheng rushed into the hospital room where I waited. She introduced herself. She had just operated on Robert. She was tiny. Couldn't have stood much over five feet. I knew the many strands of short black hair standing on her head were a result of her pulling off her surgical cap, but after the wide-eyed animated way she described what she had done—how she had dug through layers and layers of scar tissue from Robert's old hernia repair in order to reattach my ureter to his bladder—I imagined her mussed look came from her damn near needing to climb into him to get the job done.

"Why do you think he did that?" my mentor Len asked when I shared the series of events that followed the operating room decisions of Kobayashi, Robert's first transplant surgeon.

Len Syme had been a mentor of mine during my time as a Kellogg Scholar in Health Disparities, soon after I joined the faculty at Highland Hospital. He was a professor emeritus at University of California, Berkeley, and was considered the father of social epidemiology for his groundbreaking work in the social determinants of health—the concept that a person's environment determines their health outcomes more than genes, more than health behaviors, even more than medical care.

"Laziness," I said without hesitation as I looked at him.

Len looked good. Over the years I had grown to worry about him the way an adult child worries about her aging father. The last time I saw him about six months prior he seemed to have aged too much, even for someone in his eighties. But this time his hair and beard seemed less gray and more silver. His skin less pale. He had gained back some weight. He even stood taller, almost eye-to-eye with me again.

"Interesting," he said. He reminded me of how I had earlier in that same conversation attributed my and Robert's first experience with the kidney transplant system to race, with equal conviction.

"But you would have been equally justified in believing that this experience happened because of race," he added, "because when one's experience has been that bad things happen over and over again to them and the people who look like them, to attribute those events to the thing they have in common is a natural and valid response."

*Huh,* I thought. I had never made the comparison, never asked myself why I had labeled the events as I had. First I told myself that I was looking at the surgery purely through the lens of a physician, not a Black person. And as a physician it hadn't occurred to me that another physician might either deliberately or subconsciously treat a patient differently because of his race—even though centuries of health care in American history begged to differ and even though I *had* drawn that conclusion just a few months before the surgery.

Maybe the truth is I didn't want it to be about race. I didn't want it to be about something neither Robert nor I could ever change or hide about ourselves. I was tired of feeling, believing that it *is* about race.

Laziness, on the other hand, could be resolved with a nap.

But even if it had been laziness, my assumption made me wonder, had Robert been someone Kobayashi more closely identified with, if he would have made different choices in Operating Room 18 that day, nap or no nap. If he would have fought his way through the almost-impossible-to-get-through tissue if it was an Asian man lying on that table. If he would have tried a little longer to get that stent to pass across his surgical connection. If he would have started over on Robert's left side, or just started there to begin with even though it would be a little harder.

After Cheng left me, I went again to the nurses' station to ask if I could go see Robert. I was told they weren't ready for me yet in recovery. I imagined the nurse was doing the usual—getting all the paperwork done and tidying everything up before family was ushered in to see the patient, down to making sure the clean white sheet was folded neatly across his chest. I thought about the implications of the tidying as I waited and many times over as I have moved along in my career. Though I understood the tidying was meant to give the very true perception that the family's loved one was being well cared for, I believed it covered up the reality—particularly in the cases of the very old or the critically and irreversibly ill bodies trying to die in spite of our machines and drips. I often found myself wishing that we would let the family see the ugly—the blood from the procedure or oozing from their nose, the shit filling the colostomy bag, the blood-tinged snot suctioned out of the breathing machine tube shoved down their windpipe.

I say let family watch as their loved ones groan when their deep pressure sores are cleaned and packed with fresh gauze, when their breathing tubes are suctioned, or when they are just being turned for bathing. After all, these are all the things that must be done to take loving care of patients. If we let family see, then maybe there would be less willingness to hold on to a beating heart as enough life to insist upon every available intervention when no meaningful recovery is possible. If we let them see, maybe then there would be less fixation on details that don't matter, like when her toenails were last cut and when his beard was last shaven, rather than on the life being lived—or not.

In Robert's case, I knew he needed me more than I needed to see tidy. I didn't care how ugly it was, because I already knew what that looked like. And more importantly, I knew that the sooner he saw a face that loved him, the better he would feel.

Instead, Robert opened his eyes before the nurse could nudge him awake and long before I was at his side. No one was there by his side. There was only a dialysis machine and he was tethered to it. Again.

He sobbed.

He sobbed like a stoic linebacker of a man who could no longer deny his emotions. He sobbed his disappointment, his pain, his hurt, his anger.

"Did I lose the kidney?" he asked haltingly as he tried to regain his composure when the nurse appeared at his side. He needed to confirm what he thought he already knew.

"No, no, *noooo*," she said, trying to calm him. "The surgery went fine. The kidney is fine. We're just taking some fluid off of you because you gained so much."

He exhaled in relief. *As long as the kidney is OK,* he thought as he drifted back to sleep, *I can deal with the rest*—the drains hanging out of his belly, the pain of his now even longer incision, the constipation from all the pain pills that didn't work, the getting poked and prodded.

It's good that he came to that resolve, because fifteen minutes later an ultrasound technician began pressing her probe all over his fresh wound.

When they finally let me in, his tears were dried and the sheet was folded neatly across his chest—and lots of pale yellow urine poured into the collection bag.

The next day we learned that the ultrasound and a sample of the kidney taken during the surgery were normal. The worst was over. A future seemed tangible.

Robert started making plans again.

His first was to formally propose and present an engagement ring to me a couple of weeks after our hospital stays. Grown folks don't need a few years to decide if this is the right thing to do. When you know, you know—especially when life had already taught you as much as it had us.

Without my knowledge, he had called my parents to ask for my hand.

"Your dad was so pleased," he told me years later, smiling broadly. "He said no one had done that before."

I was touched to hear of Robert's deference to my parents. I was also surprised to hear he was the first to do so, since my three sisters and I had generated six sons-in-law altogether before Robert entered the picture.

I was visiting Robert's home with his parents and relaxing in an armchair when he entered the room. He gingerly got down on one knee and held open the box to reveal a lovely diamond engagement ring.

"Will you . . . marry me?" he asked softly with his voice trembling, his cheeks blushing, and his right temple twitching.

"Yes." I smiled at him, surprised at how nervous he was. *How sweet*, I thought, because there could not have been a surer thing—I had just given him my kidney. Of course I would give my hand.

# 8

## THE CANDLE IS LIT

*I hope nothing goes wrong today*, Robert thought when he opened his eyes on the morning of August 6, 2005. It was nearly four months since his complicated hospital stay and hope was yet again part of his vocabulary.

The digital clock on his nightstand read 6 a.m. Had this been a dialysis day, he would have already been tethered to the machine for nearly an hour. But it wasn't a dialysis day. He didn't have those anymore. Today was his wedding day. He climbed out of bed with purpose.

Our wedding day was among the first things Robert had planned in a long time. I had a wedding before and would have been content to elope, but Robert wanted a wedding—"This will be my only wedding," he insisted, and went about researching caterers, videographers, florists, bakers, musicians, and the like, making me the most carefree bride ever. He would present three options, any of which he would be content with, for me to choose. I was the opposite of a reality TV bridezilla, perhaps an artist's dream bride because when we sat with the florist and baker and explained the *feeling* we were going for—light . . . summery . . . less formal—I would add, "Do whatever you want to do that fits in with that." Their eyes would stretch and flitter with the excitement of being allowed artistic liberty, but also with a hint of disbelief. I would nod my certainty.

My only responsibilities were the invitations, wedding programs, and my dress. I kept the invitation and wedding program simple and pretty. However, what started out as a plan to buy a simple, inexpensive sheath of a wedding gown was quickly cast aside when the saleswoman brought out a beautiful off-white, full-skirted gown with a short train. It was strapless, fitted through the bodice, and trimmed in pink flowers and lace.

"I know this isn't what you had in mind, but I think you should just try it on," she said.

I don't consider myself the girliest of women—I stopped straightening my hair years earlier in part because I hated the beauty salon, and I preferred my Ecco slip-ons with menswear socks to Jimmy Choo heels any day—but damn if I didn't gasp at the sight of myself in the mirror in that gown. And when she brought out the matching tiara . . . I was done for.

Robert was nervous as he began to dress that morning. Just when he had begun to start wearing short-sleeved shirts again, he had his lopsided belly to contend with. His abdominal muscles had been cut twice within an eight-day period. They refused to snap back to their former shape, leaving the location of his new kidney apparent. But he began to feel a little less nervous as he admired himself in the mirror. *OK, I look good in this,* he thought—and headed happily to the wedding site.

Dunsmuir House and Gardens sits on fifty acres of the hills of Oakland, California. According to the website, the main house was built in neoclassical revival style (which sounds a lot better than the plain old Southern plantation big house that it looked like to me) in 1899 by Alexander Dunsmuir as a wedding present for his beloved fiancée, Josephine. Sadly, they never lived in the home together—he died on their honeymoon. The estate was purchased by the City of Oakland in the early 1960s and eventually was designated as a National Historic Site. The Dunsmuir staff said our wedding would be held outside on the expansive meadow, which, again, I suppose sounds more elegant than how I might have put it—the wedding ceremony near the pond, the group picture on the front steps, and reception in the front yard.

We planned the wedding of our dreams, inviting about 150 family members and friends, keeping only the traditions we wanted and adding those from other faiths that spoke to us. The wedding program would include a Donate Life plastic bracelet and information on how to become an organ donor. A quartet would play classical pieces by Bach and Beethoven through the wedding and a DJ would spin Beyoncé, Montell Jordan, and other R&B artists during the reception. There would definitely be no Electric Slide. And our first dance would be a choreographed foxtrot routine to Barry White's "*You're the First, the Last, My Everything.*" We chose a friend to read one of the sacred Hindu Upanishads, we would light the Unity Candle, and Melanie, who at that point would take the credit for bringing us together, would read a piece explaining the Black American "jumping the broom" tradition before we did it. We chose to not have a wedding party so Robert didn't have to suffer through having to decide on a best man and I didn't have to try to find a bridesmaid dress that looked good on every body shape and skin tone and had as much of a chance of being worn again as did my gown. Five-year-old Avery would walk me down the aisle, so we made the reception a kid-friendly affair. We hired babysitters for the little ones while a clown and a centipede-shaped bounce house entertained kids Avery's age. I didn't see the point of getting a limousine to use for a few hours, so I rented a minivan for the week while my family from North Carolina visited. I proudly drove that minivan up to the site with my parents, my sister Regina, niece Aisha, and Avery in tow.

Robert had hired a coordinator for the day and arrived at the wedding site three and a half hours early because he didn't trust her to make sure everything was in place and that nothing would go wrong.

But, of course, many things went wrong.

What began as a beautiful rendition of Henry Purcell's "Trumpet Tune and Air" heralding my arrival fell apart as Avery escorted me to the altar. Notes that were meant to be bold began to drizzle out as if the trumpeter had simply forgotten how to hold his lips against the mouthpiece. I tried not to register the disappointment on my face as the violinist touched the trumpeter's knee, her signal for *Just stop*. Instead I shifted my eyes to Robert's face. There I found a calm smile and a quick wink of his right eye that said *C'mon, girl, let's do this. I've got you.* My authentic self and my ego leaned back together and exhaled a smile of satisfaction that all was well.

So fixated I was on Robert's face that it wasn't until I reached the altar that I realized the minister was looking at me, trying to establish eye contact. His face was drawn in with concern, as if to say *Are you sure you're ready to do this?* I felt my face flush with embarrassment, then nodded a reassuring smile and gave him the look he was waiting for—*Yes, I'm ready.*

Reverend Noel stood tall and slender in his black robe. He had been Robert's mother's minister at her Presbyterian church for years. Ginger respected him and he her. He was no fire-and-brimstone preacher, but rather dignified in his tone, always drawing a modern-day conclusion from Bible verses in his sermons. But we soon learned that he could also *preach!* The difference between delivering a sermon and *preaching!* is akin to the difference between being able to sing and being able to *sang*. A person can sing if he can carry a tune, but if his voice moves people to tears—now that's *sanging*. Similarly, to deliver a sermon is to speak the Word of God authoritatively and knowledgeably. To *preach!* is to move souls to the point that people can't help but call out *All right now!* and *Amen!*

The preaching began sometime after the photographer slipped and fell into the pond when our mothers stood to light the side candles of the Unity Candle. Robert and I planned to each draw from the candles our mothers lit to light the center candle together, symbolizing the joining of two families into a new union—until we discovered that the wedding coordinator forgot to set up the Unity Candle by the altar. She sat in the front row smiling when Reverend Noel, Robert, and I all turned to look at her. She startled as if she suddenly remembered that she was actually there to work when the preacher took over.

"No matter," he began. "We don't *need* a Unity Candle to symbolize this union! The candle has already been *lit*!" He went on to preach about the extraordinarily special bond that Robert and I shared in the form of two healthy kidneys between us in a way that we hadn't thought to ask of him, but could not have planned better. *Amen!*

It was a wonderful day.

Given where we started, no one would have thought that Robert and I would ever meet in the first place, much less that our paths would merge so inextricably. My hometown of Spring Lake, North Carolina, was a 2,852-mile drive from Richmond, California, where Robert grew up—and most people never strayed very far away from either of these places.

The first harbinger that our paths would come together was probably my fixed delusion. I believed I had been switched at birth and that I was really supposed to be in California. I'm not sure at what age or how this belief came about—I didn't know anyone who lived in California and

I had never even been to California. It wasn't as if I could suggest it was some type of coping mechanism for a horrible existence. While I did have to pluck my own switch from the peach tree in the backyard for the occasional whooping, I felt generally happy until my somber teen years. And though free school lunches and Kmart "blue light special" clothes were the norm, I never wanted for anything I truly needed. That said, I did often feel out of place, like I didn't belong where I was since well before my teens. I didn't even think I looked like anyone in my family until I saw pictures of myself standing next to my mother at my first wedding. Our heads were tilted the same and I could see the shape of my nose was the same as hers.

Nevertheless, I readily admit, embrace even, that I'm just a country girl from *outside* city limits, where distances are measured in fields because there's no such thing as a block. The road in front of my house was dirt until I was six or seven years old and it didn't have a name that I had heard of or become a part of our address until many years later. My address growing up was Route 1, Box 358½. The ½ was because my father bought about a third of the neighbor's plot. Box 359 was my aunt's house across the road. Box 360 was our other neighbor's house a field away.

Spring Lake and its neighboring towns were the kinds of places where teenagers hoped to spend their summers picking tobacco because it paid twice as much as working in fast food. Where most planned to get married right after graduation and work at the Food Lion. Or at the tire factory. Or join the Army. Where the SATs were just a day of coloring in bubbles with a sharp number-two pencil and where the closest thing representing an Ivy League school on College Day was Fayetteville Technical Community College.

It was the summer after my junior year in high school when I discovered a path to something different. That spring, Ms. Davis called me into her office. She wasn't my guidance counselor, so I didn't know why she wanted to see me. I was nervous. Maybe Ms. Gilchrist, the Advanced Placement English teacher, had told her how she had walked in on my accurate albeit unflattering impersonation of Ms. Davis in front of all my classmates—deep, drawn-out growl, looking down her nose at us through the bifocal part of her glasses, exaggerating her already bulging eyes.

"I think you'd better take your seat now, Vanessa," Ms. Gilchrist said as she walked past me at the podium.

Yes, I was about to get a talking-to.

When I walked into her office, I found Ms. Davis, fat and deep roast coffee–colored, sitting at her brown wooden desk. She looked over her glasses at me. Papers were piled neatly in three stacks, but she held one paper toward me.

"You need to go to this summer program. Have your mother sign this paper today and bring it back to me tomorrow. Tomorrow is the last day to apply."

"Summer program? What happens there?" I had never heard of a summer program. The last summer I spent many a Saturday morning—before the morning sun rose too high and hot—bent at the waist over row after row of dew-covered leaves in our field-size garden, black soil pushing between bare toes, while I plucked the ripe butter beans and green beans from their stalks. Deddy (how we pronounced Daddy) would take some of the harvest to the machine bean sheller, while Mama and I sat snapping the long pods for green beans as we watched "the stories" on TV. We pinched off the tapered ends and tossed them in the scrap bucket as we watched the shenanigans of *All My Children*'s Erica Kane. Other days, while Mama worked the press at the dry cleaner and Deddy the mop at the veterans hospital, I cleaned the house and hung the freshly washed laundry out on the clothesline to dry, knowing it had *better* all be done by the time they got home.

"You spend six weeks at a college learning science and math," Ms. Davis said.

"But I was plannin' to work this summer." Fast food because I didn't have a tobacco connection.

"Now, Vanessa," she rumbled. "You have your whole life to work. You need to go to this program."

Even though I liked school and had only once in my life, in the seventh grade, faked being sick to get out of it (I *had* to see Michael Jackson being interviewed on *Good Morning America*—the morning after his moonwalk debut but before we had a VCR), the thought of spending my summer in school did not seem ideal. Yet, I did as I was told because that's how I was raised, and I sensed that this woman pushing me to do this even though I wasn't her responsibility meant it was important.

I don't remember any details of the science and math taught that summer at St. Andrews Presbyterian College in Laurinburg. But what I do remember is that my roommate went to a school where the students were taught to *prepare* for the SATs. And people from *universities* like Dartmouth and Yale set up booths on the program's College Day. I felt uneasy in this big world newly brought to my attention, but suddenly Spring Lake felt too small.

Still, the thought of becoming a doctor didn't even enter my mind until my oldest brother, Milton, suggested it just before I was to start my senior year. I didn't come from doctors, didn't know any doctors, and rarely saw a doctor. No need to see one when Robitussin, calamine lotion, and Vicks VapoRub cured all that ailed.

I was sixteen. He was thirty-two and, since he left home to join the Army before I was out of diapers, a relative stranger to me compared to the other four siblings between us. He was stationed in Germany but had come home for one of his rare brief visits.

"What are you planning to do after you finish high school?" he asked.

"I'm going to be a medical technologist," I said proudly. Exactly what that meant I wasn't sure. But it sounded good and no one I knew was trying to be one. My plan was the grandest goal within reach that came to my mind.

"You should go all the way and be a doctor," Milton said. "You can."

My eyes stretched wide. "Really?"

He nodded with a sincere raised brow and pulled-in lower lip as he turned his eyes away. Saying something of significance straight on, face-to-face in my family was akin to hugging hello or saying I love you. Just didn't happen. So for anyone in my family to say out loud that I was capable of such a thing was a first. It wouldn't be until I was a mother myself that my mother told me, "I always knew you were smart . . . I never had to tell you how to do something twice."

Until that brief interaction with Milton, the message had been to never think too much of myself. Twice in my teenage years uncles confessed to me, "You a pretty gal," like it hurt them to admit it. But a quick "Don't let that shit go to your head" seemed to ease their pain. *Thank you?* Still, Milton's words were all the motivation I needed to redirect my plans from becoming a medical technologist to becoming the first doctor in my family.

This was about the same time that Robert, a self-acknowledged knucklehead from Richmond, California, was deeply engaged in knucklehead behavior, though he started out innocently enough. Both his parents worked, sometimes two and three jobs each, so he spent a lot of time with his mother's parents—Grandpa and Big Mama, because she didn't want to be Grandma—until he was eight or nine years old.

Grandpa and Big Mama were a staple in the community, married more than sixty years. He was president of the church deacon board and the Richmond chapter of the NAACP and helped establish affordable housing. She sang in the church choir and ran the household, preparing a hot breakfast, lunch, and dinner for her husband *every* day. They attended church together every Sunday.

They went to the horse racetrack every week too. But little Robbie wasn't supposed to tell his parents that part. McDonald's cheeseburgers kept him quiet.

"C'mon, Seven!" Big Mama would shout.

Little Robbie, who looked like a short male version of his mother when he was young, with a twig of a frame and long wavy hair, would hop up and down and shout too. Because Big Mama's win meant payday for him too. And twenty dollars would buy a lot of bubble gum, Now and Later candy, and Orange Julius drinks—all of which Grandpa had introduced him to.

Grandpa was one of Robert's favorite people. He wasn't the only person Robert had met who went to college, but he was the first *educated* man he knew and the one everybody in the family looked to for help. Robert admired how he could watch TV, read the paper, and listen to the radio at the same time and seemingly pay attention to them all. He was also a dapper man. He wore slacks and a button-down shirt every day, even when he was gardening. Robert wanted to be like him.

But after Grandpa died and as Robert got older, Grandpa's influence was replaced with the lure of fast girls and bad boys.

"Most of us were not good boys. Even the good boys did bad things," he admitted. "We might not hurt you, but we did what we had to. We knew how to protect ourselves if we got caught."

The first time Robert was asked to sell drugs he was in junior high school. He said no, but by high school he started cutting classes regularly yet still maintained a B average. School was easy for him. He just followed the syllabus and turned in the homework—and forged his mother's signature so his absences would be excused. *Please excuse Robbie from school on this day because he had a dentist's appointment. Please excuse Robbie from school on this day because he had a doctor's appointment.* His multiple prior legitimate absences for chicken pox and allergy testing allowed his forged notes to pass without raising flags with the school. Baseball and football practices allowed him to be out late without raising flags with his parents.

The desire to have money and be like his friends eventually led him to give in to the repeated offers to start his own hustle. He began selling what the kids at school wanted—crank to the White boys, weed to everybody else—until one evening when he was standing around on the block with a dozen or so other young men and boys that everything became very serious, very quickly.

"Get on the ground! Get on the ground!" police shouted, guns drawn. Police cars swarmed in and sealed off the block.

The boys were rounded up and pushed against a wall in a line, hands over their heads and legs crossed at the ankles so they couldn't run. It was like a scene out of a movie shot in South Central Los Angeles. Robert was embarrassed. He was aware of what *they* said about boys like him. He was embarrassed that he had become that statistic.

Though he knew he was in trouble, he wasn't so much scared of what might happen next because he had no drugs on his person. He would be detained for intent to distribute, but not possession. Each boy was questioned, then pushed into a van. Those sixteen and older were taken to county jail, those under sixteen to the juvenile holding facility. Robert was fifteen.

Other boys in the holding facility were talking. "Man, they sending sixteen-year-olds to jail now for hustling. My man just got sentenced," one said, shaking his head as he stared wide-eyed at nothing on the floor.

Robert was nervous. Fifteen is close to sixteen and, in the days of the crack epidemic and the War on Drugs, the courts were up-charging, rounding ages up for juveniles.

But then he remembered his mother mentioning that she needed to take care of filing his birth certificate—which had never been filed with the City and County of San Francisco, the city where he was born. An opportunity had presented itself. Who could he call on to help him?

Big Mama.

Big Mama would be mad and she would yell at him for sure. But Big Mama would not tell his parents, at least not right away. And, most importantly, Big Mama knew people in positions to help him that she could call on. He called her to pick him up.

As they left the police station, he told her what the other boys were saying. "I need to be younger, Big Mama." He looked at her, eyes pleading.

She looked back at him with knowing eyes. "I'll help you," she said, and Robert exhaled.

The next day, Big Mama filed a birth certificate. One that said he was fourteen.

# 9

## KNUCKLEHEADS AND NERDS

As traumatic as the near prison experience was for Robert—enough for him to drop his hustle for good—it was not the most pivotal one in his teenage years. The most pivotal experience involved a girl he met in the spring of his sophomore year of high school. It would shape the man he would become.

Jennifer was a preacher's kid and a year ahead of Robert in school. She was the one in a clique of giggling girls who was bold enough to start talking to him. He wasn't particularly drawn to her, but he knew a couple of the older boys she had dated.

"If you get with her she'll fuck," they told him.

To dismiss a girl who clearly liked you after receiving that kind of information was to be a punk.

"Oh, you scared?"

"You still a virgin?"

These were the taunts from boys and girls alike that pressured him into his first sexual encounter at age twelve. To say no, to walk away, was to have your budding manhood called into question, you would be considered a punk. By boys and girls alike. Everybody talked. Robert had learned not to walk away, to always oblige a girl offering sex. He could think of nothing worse than being a punk. So when Jennifer persisted, he responded in kind. Within the month, her past boyfriends' prediction proved true.

They dated the rest of that school year and into the summer, longer than he had dated any girl before, so despite his multiple past encounters, it was with Jennifer that Robert learned about foreplay in order to get the response he wanted. He liked Jennifer, but by July, the relationship had lasted as long as he believed a high school relationship should.

"I think it's time for us to move on and start seeing other people," he said when she came to his home one afternoon.

"Why?" she protested, and started to cry.

He consoled her.

The following March, as a result of that consolation, Jennifer gave birth to a baby boy named Cameron.

Robert felt the gravity of Cameron coming into the world and a strong sense of responsibility for him—but not for his mother. Jennifer, on the other hand, wanted to be a family. She and Cameron were a package deal, she told him. So Robert continued to make mistakes—he tried to be a family when his heart wasn't in it; he told her he loved her when he didn't; he even gave her one of Big Mama's old rings as a promise ring for marriage someday, when he was spending time with other girls. When he could no longer pretend, she became angry. Cameron became her leverage to make Robert be with her. When he wouldn't, she became angrier. She and Cameron were a package deal, she told him again. Eventually, Robert's involvement in his son's life was reduced to financial. His greatest regret was not having the presence of mind and maturity to negotiate a different outcome. One that included knowing his son and his son knowing him.

Had I met Robert when we were teenagers, I doubt there would have been any falling in love or kidney donating in our future. His world was foreign to me. A boy cutting school would not have appealed to me. Sneaking out of the house at night, like I learned my sisters did only after I was grown, did not occur to me. And while I was not a virgin on my wedding night, my deflowering was much later than Robert's. I didn't even take my first drink until I was twenty-one. I was a good girl, a nerd.

However, I must admit that the Richmond in Robert appealed to me, the grown-ass woman, when we did start dating. Perhaps because the Richmond was wrapped up in his metrosexual-ness—his fashion sense, the love of mani-pedis, facials, and really anything "spa"—and *his* nerd, the thuggish tendencies seemed to complete the picture. Equal parts nerd, metrosexual, and thug. What every straight girl *needs* in her life, because that's a man who will always have a job, who you won't have to dress for any event, and who would know how to protect you if needed.

While Robert was being a knucklehead, I was daydreaming about being a doctor, about "helping people." As I headed off to Duke in Durham, North Carolina, for college, I saw myself becoming a country doc, tending to whatever ailed. Sore knees. A cough. High blood pressure. But like a fickle college kid changing majors with the seasons, once I started medical school, becoming an obstetrician/gynecologist was the goal. How I came to that decision I can no longer remember.

Duke had an unusual curriculum in that by eliminating summers off, the third year was time designated for earning a master's in public health or completing a research project anywhere in the world the student could find a legitimate and willing mentor. The first year was all lectures, the second year required clinical rotations, and the fourth and final year was clinical rotations of the student's choosing. For my third year, I found a mentor willing to work with me on a gynecology-related project at the University of California, San Francisco. I don't even remember considering anywhere else.

In the summer of 1995, my country-girl self with no sense of direction piled into my little burgundy Geo Metro with a couple of suitcases, a United States map, and my first cell phone—a huge apparatus that had its own black canvas shoulder bag and was so expensive that I intended to use it only in case of emergency—and headed to San Francisco alone. I took the northern route via Interstate 80. I approached it like it was a job. I drove all day and stayed in modest hotels at night, where I bought postcards to mail to my mother on my way out the next morning. It took me five days to get there. I stared in awe at the Bay Bridge as I drove across it. My heart quickened with excitement and anticipation of what was to come.

That was the summer after the one Ginger spent in the hospital, when Robert, knucklehead no more, spent the months afterward in a labor of love. It was a demonstration of how close he felt to his mother.

But their relationship wasn't always that way.

He remembered a time when he was mean to her, because she had done things to anger his seven- or eight-year-old sensibilities. A Mother's Day card that initially read *I love you* and *You're the best mom ever* was revised to *I hate you* and *You're the worst mom ever.* Another time there were tacks placed strategically under covers on Ginger's side of the bed.

But he also remembered her giving it back in spades during that time. For stealing ten dollars from her purse, his consequence was her complete silence. He couldn't even get tucked in or get read a bedtime story—for two months. It hurt him to his core.

It was years into our marriage before I realized that it was this experience that made him immune to my comparatively sad attempts to give him the silent treatment. Before twenty-four hours passed, I would be the one writhing in my own silence while Robert was just settling down into another good book as if to say *Twenty-four hours? Girl, please. You gotta do better than that.*

By the time Robert entered high school, his relationship with Ginger had shifted. They had more time alone together, and he realized he liked her. There were long talks. She talked to him about her relationships with her parents, her siblings . . . his father. He came to her and told her about getting Jennifer pregnant while his father heard about it from somebody in the neighborhood. It was her words, *"I'm so disappointed,"* every time he behaved very badly that hurt him more than any punishment ever could.

*I'm so disappointed.* The words came with a slow shake of the head and a deep sigh.

It was these words that led to his decision to go to Morehouse College. He knew that if he stayed at home he would continue to do things to make her say them again. So because she spoke many times of four men she had met and admired—Martin Luther King Jr., whom she met once in Los Angeles; Richard Caesar, a dentist in San Francisco and the son of his namesake, Robert Louis Caesar, who was an educator and like a second father to her; Benjamin Elijah Mays, the seventh president of Morehouse College; and Howard Thurman, who pastored the Church for the Fellowship of All Peoples, the first integrated interfaith religious congregation in the United States—who were all Morehouse men, Morehouse was where Robert wanted to go. He wanted to be like the men Ginger admired.

By the time I had to go back to Durham to complete my fourth and final year of medical school, I had become seriously involved with Avery's father, so my return to California was already planned. As I rotated through a new specialty every month of that last year of medical school, I learned that each one attracted a fairly distinct personality. It is a certain type of young doctor who is attracted to pediatrics. She can engage children in a way that eludes the internist. It is a very different type who is attracted to surgery—

she wants to fix things. Still another to anesthesiology—she gets a thrill out of putting people to sleep. And yet another to internal medicine—making the diagnosis turns her on.

I learned my personality did not jibe well with ob-gyn. While I loved the women's health and outpatient clinic aspect of the field, I hated the hospital side. My love-hate attitude clearly showed on my face and in my demeanor. My evaluations from clinic doctors were glowing. Those from the hospital doctors? Not so glowing. The doctors I was exposed to on the hospital side were wonderful with their patients but always seemed angry and sleep-deprived. They also wore a bit of a chip on their shoulder, a chip that seemed to leave them trying to prove to the general surgeons that they were real surgeons too—by venturing outside the pelvis to remove appendixes and gallbladders for no other reason than it might cause problems for the patient later. Occasionally the wrong tube would be cut, resulting in a long, complicated hospital course for the patient and reinforcement of the negative stereotype of the gynecologic surgeon.

Mostly I knew ob-gyn was not for me because I couldn't get through the surgeries. After a couple of hours behind the surgical mask and under the hot surgical gown, performing the medical student's primary task of pulling back on the metal retractor that held the skin, fat, and muscle layers so the surgeon's field of operation remained in constant, unwavering view, the light in the room would fade for me. I would have to ask to sit down or risk passing out face-first into the surgical field. Looking like a wimp who couldn't even stay upright for a few hours was bad. Contaminating the patient would have been far worse. Perhaps internal medicine would be a better fit for me, I thought.

I ended up in a primary care–focused internal medicine residency at Highland Hospital in Oakland, California, where Robert happened to serve as treasurer of the board of trustees.

So entranced by my involvement with Robert, I clearly forgot about the "every specialty has a personality" lesson when I made the decision to become a nephrologist. Unlike the usual nephrologist in the making, I was not fascinated with everything kidney by the end of my intern year. Early in my second year, as a resident, I did not seek out patients admitted for anything kidney-related. I did not use what I had learned from various cases of kidney problems to impress nephrologists and I did not ask to work the renal service at two or three hospitals late in my second year. I did not spend months of my third and final year of residency in a black skirt suit and sensible shoes traveling from city to city, hoping to convince the best nephrology training programs to offer me a fellowship position.

Rather, I was fascinated with how going through the process of becoming Robert's kidney donor gave me a glimpse of the kidney transplant system that being a primary care doctor did not provide. Though at the time I was working on research projects on the effects of language barriers on health outcomes, my experience with Robert inspired me to change my research focus to what made some people much more likely to get a kidney transplant than others. Donating my own kidney was my solution for Robert. I saw research as the way to help other people like him.

I arranged to meet with nephrologist Dr. Michael Chort to get started on my new research interest. Dr. Chort's pale face was pleasant without effort. He reminded me of Humpty Dumpty with a receding hairline, but his status as master clinician and internationally renowned researcher made him a rock star in the world of academic medicine. I was honored he agreed to meet with me.

"You should become a nephrologist," he said, his clasped hands resting on the conference table between us. "I can't tell you how many journal articles I read by non-nephrologists and think, 'They just don't get it,'" he said with a wave of his hand. He might as well have been wearing a T-shirt that said IT'S A NEPHROLOGY THING. YOU WOULDN'T UNDERSTAND. "For just one tough clinical year, you could really strengthen your research depth and perspective . . . and how your work would be received."

It hadn't occurred to me to specialize in anything, let alone nephrology. Nephrology was considered one of the most, if not *the* most intellectually difficult of specialties. However, the notion that becoming a nephrologist would get me and my work taken more seriously resonated with me. For all I knew, Chort's words *You should become a nephrologist* may just have been the generic recruitment pitch he delivered to anyone interested in kidney research, but I took them as a vote of confidence that *I,* the little Black girl from Spring Lake, North Carolina, *could* be a nephrologist. A request and a couple of interviews later, I was accepted to begin the UCSF nephrology fellowship in July 2007.

# PART III

## SHE IS A THING OF BEAUTY

# 10

## HER

It was a Friday night in a week of sixteen-plus-hour workdays and the work was done. All my patients had been seen and examined. All the labs checked. All the notes written and all the recommendations communicated. My shoulders ached. My feet hurt. My *eyeballs* hurt and the edges of my brain felt blurry.

But it was no TGIF moment for me. It was July 2008 and my first weekend on call as a renal fellow at the University of California, San Francisco—the hospital of the often rich, sometimes famous, and usually complicated people of an entire region. It was a demographic that frequently created high levels of anxiety for the doctors in training and an abnormally high anal sphincter tone among administrators and attending physicians, making UCSF the most demanding of the three hospitals we rotated through as fellows.

This weekend would be spent seeing all of my patients and those of the two other fellows so that we each would only be on call every third night and have one weekend of the month completely free of the hospital. I would find out how much fluid was pumped into the patient's bodies in the last day. How much pee had come out. What their lungs sounded like. How their lab values had changed since they were last checked four hours, eight hours, twelve hours earlier, and discuss it all with the attending along with my thoughts on how their care should proceed for the next several hours. I would see every new patient any other medical team asked for nephrology input on. If a patient came in at any time during the night that any doctor in the hospital thought *might* need dialysis, I would return to the hospital for the minimum of two hours it would take to review the medical chart, examine the patient, talk through the case with my attending, communicate recommendations to the team requesting the nephrology consultation, and leave a note on the chart. And since even the most empathic of us devolve into a basic selfish human in the face of severe exhaustion and sleep deprivation, I would be thankful that my to-do list in the middle of the night would not include talking to the patient, as most would be comatose. Having to talk to the patient would add a half hour or so. A half hour or so that I wanted to be asleep.

So the work was only done for the moment. I hoped, no, *prayed,* the work would stay done long enough for me to shovel some dinner into my face and get a few hours of uninterrupted sleep before the piercing *beeeeep, beeeeep, beeeeep* of my pager began again, demanding my attention.

I walked to Fourth and Irving, but there was no seafoam green Prius. *Don't panic,* I told myself. Maybe it was Fifth and Irving? No, no seafoam Prius there. Maybe I just didn't see it before. I walked back to where I thought I had left my car. Then to where I might have left my car. Sixth and Irving? How about closer to Hugo? Or maybe as far down as Seventh? Then back again. And again. For an hour. Feet throbbing with every step. Where was my car? Had it been stolen? Tears began to roll down my cheeks.

I called Robert.

"I can't find my fucking car. I don't know if I'm cut out for this." Not that I had never forgotten where I had parked my car in the past, but in that moment with my sore feet and eyeballs, losing the car was about so much more than the car. It was a metaphor for my life. Who did I think I was anyway? I had no business venturing into this strange land, so far away from Spring Lake, North Carolina, I told myself. I wanted to retreat.

"No, I wouldn't say that. It's going to be OK," said Robert, his baritone voice almost purring. Like he believed it really was going to be OK. I almost believed him.

"Call the police," he said.

"OK," I sniffled, happy to have someone to do the thinking for me. Happy to have Robert to balance me. I've always admired his evenness. No dramatic swings from feeling like the world was tolerable to equating a missing car with the meaning of life. I have often wondered how much of the evenness was just Robert and how much of it was an adaptation to living with a serious illness. One that had the power to end his life. At any time. Without warning. I imagine it helped him put things into perspective.

From the police based at UCSF I learned that my car had been towed. My rear bumper had jutted several inches into someone's driveway and that someone was teaching me a lesson. *Just one tough clinical year*, Dr. Chort's words came to mind. Now they sounded like the painfully understated warning of the doctor, poised with a huge needle in hand: *This might pinch a bit.* Can't say I wasn't warned. But if it had rained, or if I didn't have the three hundred dollars to get my car out of the impound, or if my pager had sounded before I could get home and get a nap—just one more insult that I would construe as yet another sign from God and the entire universe that this was not the path I was supposed to be on—I might not be a nephrologist today. Perhaps too much for even Robert to reel me in from. But no downpour matched the tears streaming down my cheeks, my bank account still had a few hundred dollars in it, and I got to sleep three whole hours before my pager went off, so I stuck through that one tough year, entrenching myself in everything kidney for the first time in my medical training.

As a medical student a decade earlier, stroking brains, failing hearts, phlegmy lungs, inflamed pancreases, and even diarrheal colons dominated my experience of the hospital wards.

The kidneys didn't win my attention. And because I could graduate without spending a month on the renal service focused only on dysfunctional kidneys, that's exactly what I did.

But after medical school, there is residency—and no one escapes renal in an internal medicine residency, especially at Alameda County Medical Center, Highland Campus, in Oakland, California, where I trained. While Highland was affiliated with the prestigious University of California, San Francisco, we knew that our large number of foreign medical graduates made us internal medicine residents the bald-headed notch below redheaded stepchildren of UCSF. At Highland we didn't even have fellows around who were specializing in anything—so everything down to subspecialty level calls in the middle of the night fell directly to the resident.

I remember being on call my first night as a resident on the renal service. Only the team accepting patients being admitted to the hospital spent the night there, so when my pager went off shortly after 1 a.m. I was in my own bed at home. I cringed when the beeping pager startled me awake.

I called the number back and was told about a man who drank antifreeze. The kind with ethylene glycol in it. Besides children and animals who sometimes drink antifreeze accidentally, I knew alcoholics who couldn't get their hands on anything else would sometimes resort to drinking antifreeze because the ethylene glycol made it sweet. I also knew antifreeze could kill a person quickly, not from the ethylene glycol itself, but from the glycolic and oxalic acids the body breaks it down into. It is these that wreak all kinds of havoc—shutting down brain, heart, and kidneys. The patient

I was hearing about had already gone from intoxicated to unconscious. Not a good sign.

While the medication fomepizole prevents ethylene glycol from being broken down, it is very expensive and doesn't just come off the shelf in places like Highland Hospital. It would take hours to get it, and my new patient didn't have hours. Hopefully he had bought himself some time by getting drunk first, as ethanol, the poor man's antidote, also prevents ethylene glycol from being broken down. But hemodialysis could filter away the ethylene glycol and all the ensuing badness from his blood and save his life—if he got it quickly enough. It was my job to get to the hospital and get the patient started on dialysis before it was too late.

But I didn't know how to put in a dialysis catheter, much less how to write dialysis orders. The bore of a dialysis catheter is like the thickness of a pencil. A simple IV wouldn't do. I could feel my heart racing, pounding, and my armpits getting sticky.

I hung up and called the renal attending, Dr. Ling, the supervising doctor scheduled to be on call with me, and explained the situation. The patient's blood showed a high anion gap, a large difference between positively charged sodium and negatively charged chloride and bicarbonate. The difference, the gap, is larger when there are a lot of other negatively charged substances in the blood—like the acids ethylene glycol is broken down into. He sighed. Like he really didn't want to be bothered. He sighed again when I told him I didn't know how to put in a line. Like he suddenly realized he couldn't just tell me what to do from his bed. He would have to come in to the hospital too.

"What is the osmolar gap?" he asked. "Because if the gap is closed, it's already too late."

My eyes shifted left from the darkness before me to stare in disbelief at the receiver. *Are you really trying to find a way out of this situation?* I thought.

But what I said was, "*Um,*" feeling very stupid because not only did I not know the answer to his question (the *least* a doctor in training can do is gather the necessary data), but I also didn't know that a closed osmolar gap—no difference between how many dissolved substances we calculated were in the blood versus those measured—meant all the ethylene glycol had been broken down and the damage was done.

"I don't know," I admitted.

I called the ER back. Those results weren't back yet and the patient was being moved to the ICU. Weren't we wasting time? I called the attending back. He sighed again, but agreed to meet me at the hospital.

It took me fifteen minutes to get there. I hurried in through the entrance at the back of the building adjacent to the employee parking lot, my heart pounding beneath by scrubs and white coat, my pager, stethoscope, penlight, reflex hammer, keys, a few pens, three pocket-size medical quick-reference guides, and a dozen or so large index cards each scribbled with the details of my patients weighing me down. Squares of linoleum rushed by beneath my sneakered feet. A few yards ahead a tall, older Black man in dark brown coveralls was mopping the floor in front of the elevators, a big yellow bucket with wheels and a CAREFUL WET FLOOR sign propped nearby. I paused, looking for a dry path to tiptoe across.

He stopped mopping and looked at me, one eye squinted.

"*Nah-uhn,*" he said. "You need to go 'round to the other side of the building," as if keeping the floors clean was the most important activity of the hospital, trumping why I was running.

"I'm sorry, sir, but I'm trying to keep someone from dying!"

He brought the mop handle to his chest, conceding to my made-for-TV drama if nothing else, and I did my best not to leave footprints on his freshly mopped floor. The elevator ride to the fourth floor was quick, in my mind sped up by my incessant pushing of the brightly lit 4 button.

In the ICU, I gathered the supplies to put in the dialysis catheter. Catheter kit. Sterile gowns. Sterile gloves. Protective face shields. A dialysis nurse had been summoned and was setting up the machine and awaiting her orders. Dr. Ling arrived shortly thereafter. He was a good-looking Chinese man. Tall, slim, muscular, in his mid-thirties. He was part of the private nephrology group that the hospital contracted to provide nephrology services. I hadn't met him before.

He fumbled with putting in the line as if it had been a while since his last one. Hard sighs, but few words. We didn't talk much. Looking back, the few minutes after getting the line in and the patient on dialysis would have been an incredible opportunity for Dr. Ling and me to take a sample of urine to the lab's microscope to see the envelope-shaped oxalate crystals in real live urine, not just in pictures. Maybe even break out the black light of the Wood's lamp to see if the urine would be fluorescent under it like the books said it would. This is the kind of sexy stuff that attracts medical students and residents to nephrology.

But Dr. Ling didn't even ask if the patient was still making any urine. It was sometime after 2 a.m. and it seemed to me that he was not interested in teaching or recruiting a scared little resident who didn't even know to ask the ER doctor if the serum osmolarity had been measured. My impression was that he just wanted to get the work done and get back to his bed.

We were too late. The patient died before morning.

Perhaps if the patient hadn't died or if Dr. Ling had seemed the slightest bit interested in teaching me, I would have been drawn to the sexy side of nephrology in that moment. But the patient did die and Dr. Ling wasn't interested, so I wasn't drawn. Nephrology made me feel stupid and useless, yet years later there I was starting a nephrology fellowship with six other wannabe nephrologists. They were all straight out of their residency programs, where they had no doubt focused on everything kidney for at least a year or two, while I had spent a couple of years as an internal medicine attending at Highland followed by two years as a general medicine clinical research fellow with no desire to focus on anything kidney. So by the time I reached the nephrology fellowship, I felt even dumber about the kidney, but maybe slightly less useless because at least my new kidney-focused research would help somebody someday, I hoped.

That I started with the research year rather than the tough clinical year fellows usually started with didn't help my confidence. By the time I expressed interest in the fellowship, the program had already filled its clinical positions. I could have postponed a year, but I was eager to get on with my research career.

My decision not to postpone a year turned out to be a poor one because the research fellow takes over the responsibilities of clinical fellows on vacation. Since the research fellow is usually in her second year, she already knows a lot of nephrology and is almost ready to function independently, without attending nephrologist supervision. I could not have been further from functioning independently. This not only undermined my confidence in nephrology but also negatively affected my first impression on attendings. I felt I was in the way, like the attendings' time spent taking care of patients would be much more efficient if the fellows—especially the ones who weren't all about nephrology—would just step aside.

The fact that I was the only Black person in the entire division of nephrology across the program's three hospitals didn't help my confidence. When a person sees no one who looks like them, they question if they belong. It was the reason becoming a doctor hadn't crossed my mind until my brother suggested it.

I remember getting my first hint that I was the only one like me on the first day of fellowship. That first day was to be filled with paperwork and the start of the crash course on how to be a renal fellow at the University of California, San Francisco. I wasn't late, but I was the last to walk in and felt very late to the party because they were all nearly a decade younger than I was. A large white rectangular table filled most of the small conference room. My new colleagues sat quietly around it, already busily filling out their forms. All but one had a brown face, but none like mine. One Chinese man. One Indian man. Three Indian women. Only the fifth brown face looked up as I entered. It belonged to a plump Latina. She *beamed*. Her eyes stretched wide and her mouth even wider to show upper *and* lower teeth. I understood her joy in seeing a brown face more like hers. It dulled the feeling of oddness, of aloneness. I nodded back a close-lipped smile and sat in the empty chair at the end of the table where my paperwork awaited.

I worked steadily on my research project and did the best I knew how to do when on clinical service, but it would be months into the fellowship before I gained any real attraction to the kidney. After that first day of paperwork, we fellows would continue to gather around the conference room table every Tuesday morning at 7:45 for the fellows lecture. A lecture on everything kidney in forty-five-minute chunks. Electrolyte disorders. Acute kidney injury. Renal causes of hypertension. They tended to be PowerPoint presentations, from super-smart faculty who were often less than super at captivating me with the detailed information littering each of their slides.

I remember one lecture that did capture my attention. Renal pathology. It was at the fingertips of Dr. Sipos, a small, white-haired pathologist who stretched his mouth purposely so that each European-accented word was clear, punctuated English. With a click on the computer, his finger summoned a gray-scale scanning electron micrograph of the normal glomerulus, the kidneys' filtering unit, magnified eight hundred times then projected onto the screen.

I was mesmerized by the beauty in its design. The glomerulus was stripped of its cap and looked like a ball of tightly coiled yarn. Each strand like interlaced fingers. Hundreds of them, all emanating from the large lumps of podocyte cells scattered across the surface. The podocyte was like a mama octopus trying to hold all her babies at once, each tentacle with a set of fingers and each finger with another set of fingers, until all the fingers of all the podocytes clasped and touched every curve of the tuft at once. So tight the clasp, only the tiniest bits could slide through, only to be captured by the ill-fitting cap of the glomerulus.

The cap funneled these liquid bits into the tubule, a half-untwisted paper clip maze of a tiny tube. Together the glomerulus and its tubule made up the nephron. Each tubule lined with a single layer of different kinds of cells along its length, each type of cell distinct in appearance and the work it did. I imagined each cell was like a factory worker stationed at a fluid conveyor belt. I imagined the workers of the proximal tubule as muscle-bound, doing the bulk of the lifting, removing almost all of the blocks of amino acid and glucose and a large portion of the small blocks of salt, potassium, water, phosphorus, and bicarbonate from the fluid conveyor belt while placing medication leftovers and wastes on it. Next, in the still-twisted part of the paper clip, the Loop of Henle, the workers selectively pulling salt or water from the conveyor belt as it dove deep into the kidney and back. I imagined the workers near the top bulked up in rain gear; no water slipped by them. The workers in the last untwisted part of the paper clip, the distal nephron, were supervised by the hormones aldosterone and vasopressin, which made sure workers pulled or added blocks of potassium, salt, and water as ordered to fulfill the needs of the rest of the body. Finally a set of workers interacted with the fluid so that the body's excess acid was released. What was left on the fluid conveyor belt was passed into its collecting duct, which met other ducts to eventually leave the kidney through the largest tube, the ureter, funneling the end product, urine, into the bladder. What was removed from the fluid conveyor belt was returned to the tiny capillary blood vessels that wrapped around it and the workers.

Dizzying it was to think about how this glomerulus bobbing in its cap funneling into its tubule all wrapped in its capillaries was just one filtering unit. One nephron. How in life it was repeated two million times over—one million in each normal kidney. I tried to imagine the capillaries tying it all together. Two webs of them, coursing through every glomerulus and around every tubule. One receiving blood, the other ushering it out. Like a blood vessel Autobahn.

Each web merged into larger vessels, then still larger ones, then even larger ones, then larger ones still—until one web of capillaries of all the nephrons of each kidney met at the renal artery branching directly off the aorta, the largest artery in the body, pushing about a half cup of blood into the kidney every minute of every hour of every day. And the other web of capillaries of all the nephrons in each kidney met at the renal vein situated just below the renal artery and branching directly off the vena cava, the largest vein in the body, running alongside the aorta and returning the blood, almost all protein, and everything filtered through the glomerulus but removed from the conveyor belt to be put back into the body's circulation.

One million nephrons, all packed into a bean-shaped sheath less than five inches long and weighing only about a third of a pound. Times two. Each kidney wrapped in a padding of fat and tucked beneath the ribs and muscles of the lower back, on either side of the spine, working as one unit. My thoughts spun with the microscopic complexity and precision of it all. Yet there was still so much more left to wrap my brain around. Like how the juxtaglomerular apparatus included a clump of specialized cells beside the glomerulus entry that produced renin, an enzyme that activated a hormone that controlled blood flow into the glomerulus by making blood vessels smaller, and another clump of specialized cells that connected the thick part of the paper clip loop to its glomerulus and controlled the speed of the conveyor belt based on how much salt was going by. Like how the early tubule cells converted regular vitamin D into an active version that helped to regulate the body's calcium and phosphorus. And like how the cells between nephrons made the hormone erythropoietin, which prompted the bone marrow to make blood. As if the production of urine wasn't enough. It was in that moment that I began to appreciate the kidney and all it did. I was in awe.

My attention zoomed back to the image on the screen. How beautiful it was. Intricate. Curved. Complex. Multitasking. Like a woman, I thought. Nothing this beautiful is just an *it*.

The kidney. *She* is a thing of beauty.

But anything so beautiful and so complicated is bound to get hurt. And so She does in many ways.

# 11

## BABIES

There's no good time for a woman doctor to have a baby. Men doctors have wives to do it for them. But neither all-night studying nor hundred-hour weeks on the hospital ward is good for pregnancy. Babies sometimes get too sick for day care. And peers tend not to love taking extra call nights because somebody decided to try to have a life outside the hospital.

I got pregnant with Avery *on purpose* as an intern. While my program director was wonderful, attempting to adjust my schedule to be the best for my pregnancy with the least impact on my peers, someone had to pick up the baton when I walked out of the ER in the middle of an on-call day because my baby was sick. I hoped to not repeat history with the fellowship as much as possible by at least getting pregnancy out of the way before July 2007 rolled around.

But when pregnancy didn't happen in our first two years of marriage and I was well into my fellowship, we began to worry something was wrong. I was pregnant with Avery two months after stopping birth control pills. Granted, I was only twenty-eight years old then, so there could be my advanced maternal age of thirty-seven to blame for our troubles, but Robert and I had other cards stacked against us too.

Men with kidney failure are often infertile. Sperm can be defective and low in number. Transplant restores fertility, but in Dr. Cheng's efforts to save the kidney transplant, she accidentally cut Robert's right vas deferens, the duct responsible for allowing sperm to move from his right testicle to his penis, so only one of Robert's testicles was contributing to our baby-making efforts.

Women with chronic kidney disease have issues with infertility too. They often have irregular menstrual periods and don't ovulate. As the kidney disease worsens, so does the infertility. It is rare for a woman with advanced chronic kidney disease to get pregnant. And it's even rarer for one with end-stage kidney disease on dialysis to get pregnant. When it does happen, the women often have severe high blood pressure and only about half of the babies survive. The babies that do survive are usually born two months early—in spite of the patient enduring dialysis six times a week instead of the usual three. After transplant, pregnancy is much more common, but only at about a third of the rate of women without kidney issues.

I still had one normal kidney, so I fully expected to get

pregnant. We went to an infertility clinic to find out why it wasn't happening.

Marguerite Cyders appeared to be in her early fifties. She had a kind smile that you could see in her eyes and a reputation for knowing what she was and had been doing for the past couple of decades. She laid out a clear plan for us. Her plan involved dozens and dozens of injectable medications that our insurance wouldn't pay for, so money that was being saved for a house was redirected to making a baby instead.

After weeks upon weeks of shots into my belly fat and booty, she retrieved eggs from my advanced-maternal-age ovaries. Robert's swimmers numbered on the low end of average and looked normal, but Cyders didn't want to turkey baster them into my uterus and risk the added burden of a twin (or more) pregnancy on my one kidney. The lab combined my eggs with Robert's sperm to make us a total of nine embryos. She planned to place two of the four-celled embryos inside my uterus to start, freezing the rest.

Moments before the procedure, she showed us a picture of them. It was love at first sight. These were our babies. I wondered who each might be. Which would have Robert's eyes. Which would have mine. His love of reading. My sense of humor.

But a couple of weeks later I got my period. Neither embryo had attached to the walls of my uterus. Cyders's eyes mirrored the disappointment in ours. Sometimes it just didn't work, she said. All we could do was try again. We had seven more embryos.

After another round of shots for weeks to get my uterus ready for pregnancy, Cyders tried placing three embryos. A couple more weeks later I got my period again. And again she said all we could do was try again. She placed the remaining four after yet another round of shots. The risk of a multiple pregnancy had proven far less an issue than getting pregnant at all. But my period came. Again.

And my soul ached.

We had gone through our savings and had no baby to show for it. But even if our money had no limit, there would be no more IVF. I could no longer bear the painful shots only to be left heartbroken weeks later. If we got pregnant, it would have to be on our own. Unlikely, we knew, but still we tried and hoped.

"You should be happy, you already have one baby," one of my nephrology fellow peers said one day as I jokingly whined about *feeling* my advanced-maternal-age eggs dying. *Oh, there goes another one!* She might as well have hissed, "Stop being greedy—*it ain't like it's kidney failure,*" to help me regain my perspective.

But she was missing the point. I didn't just want another baby. I wanted a do-over. I wanted to have a baby with people who loved me around. I wanted to make a baby with someone who loved me. I wanted a baby with My Robert.

Robert wanted a do-over too. A chance to have a baby with a woman he loved. A woman he wanted to be a family with.

"We found each other too late," Robert said, trying to ease my pain. And his.

It has been years, but part of me has yet to come to terms with the fact that my dream of a little combination of Robert and me would never come to be. I know it ain't kidney failure, but it feels like it in the sense that I get a monthly reminder that a part of my body has failed me.

# 12

## ZEBRAS

It was the late 1940s when Dr. Theodore Woodward, a professor at the University of Maryland, first said to his medical interns, "When you hear hoofbeats, think horses, not zebras," because in Maryland one was much more likely to see a horse than a zebra. Most of adult nephrology are horses—chronic kidney disease, with two-thirds of it caused by high blood pressure and diabetes. But it is the exotic, relatively uncommon things that can go wrong with kidneys—the zebras—that bring young doctors to nephrology. The zebras make the kidney sexy.

It was a zebra that brought Robert into my life.

It was the beginning of the summer before sixteen-year-old Robert's junior year in high school—a few weeks before he learned he would soon be going to the pediatrician with his own child—when Robert sat with his mother in the waiting room of his pediatrician's office. He needed a routine sports physical. Football season was about to begin.

Robert had been going to Dr. Philipe Santoyo's solo private practice since he was eight years old when he developed allergies to his cat, his guinea pigs, pollen, and grass. At least that's what the reactions to some 150 potential triggers injected just under his skin up and down his back had revealed. Weekly allergy shots followed until he was fourteen.

"Robert Phillips," the nurse called from the swinging door by the front desk less than five minutes after they sat down.

Robert and Ginger quickly rose from their seats and followed her through the swinging door and down the hall. They paused at the scale, where she checked Robert's height and weight. A growth spurt in his thirteenth year that left him eight inches taller and ninety pounds heavier meant he was no longer that twig of a boy who frequented the racetrack with his grandparents.

"Come on in and have a seat," she said as she ushered them into exam room 2. "The doctor will be right with you."

Ginger took a seat in the chair and Robert hoisted himself onto the white paper stretched across the exam table.

Dr. Santoyo strolled in almost immediately, Robert's clinic chart in hand. He was short, not much taller than Robert *before* his growth spurt. He shook Robert's hand firmly, then turned to Ginger.

"Hello, Mother," he said flatly in a thick Castilian accent. He was nice enough, Robert thought, but he had never known him to waste time on small talk. Ginger smiled and nodded hello.

"What are you here for today?" he asked, and Ginger handed him the sports physical form.

He took the form and a seat on the rolling stool and crossed his legs. Robert, already a budding metrosexual, noticed that the doctor was wearing his usual sports coat, shirt, and tie under his unbuttoned white lab coat with slacks and expensive loafers. This day he wore Ferragamo loafers. A sad salt-and-pepper comb-over made him just shy of dapper.

"OK, let's go through this," he said, looking at the form through the glasses resting on the tip of his nose.

After Robert's reflexes were tested, his lungs heard, and his finger pricked, Santoyo handed him a clear plastic cup. "Pee in this cup and leave the sample on the tray in the bathroom."

Robert walked to the bathroom to do as he was told. A layer of foam floated on top of his yellow urine sample. It looked like a freshly poured beer, but he thought nothing of it; other boys he knew had said theirs bubbled up in the toilet some as they peed too. His bubbles just never went down. He left the sample on the tray. A basic urine screening test was a standard part of sports physicals, and Robert had been peeing in cups every year for years for this purpose.

The nurse retrieved the sample with latex-gloved hands and placed it in a plastic bag to go to the lab. In the lab, the technician would take from the dipstick bottle one narrow strip of lightweight cardboard, all but the top inch stippled with ten evenly spaced squares, each embedded with a specific chemical that would change its color if a particular substance, like blood or protein, was present in the urine. She would submerge all the squares into Robert's pee, just for a moment, then lay it on the paper towel on the counter, giving any chemical reactions time to occur. She would come back to it in two minutes to see if any of the little squares had changed colors. One had.

Two days later, Santoyo called Ginger and said he needed to see them again.

"What's wrong?" Robert asked as he walked into the doctor's office, after Santoyo had spoken with his mother alone. He was more alarmed about meeting in his office and not an exam room than he was about the doctor talking to his mother first.

"We found a lot of protein in your urine," he said, then looked at Ginger. "Mom, I want to send him to a nephrologist."

This was the first time he had heard that word. He rolled the syllables around in his mind. *Ne-phro-lo-gist.* This didn't sound good.

"A nephrologist is a kidney doctor," Santoyo explained. "The kidneys make urine. I just want him to run some tests to find out what's going on."

Robert was aware of his mother watching him.

On their way out, nephrologist's office number in hand, Ginger asked Robert what he was thinking. He shrugged, feeling like he didn't know enough to feel any particular way about the news. He looked at her to gauge how worried he should be. If she was worried for him, it didn't show on her face. This was just some more tests and doctor visits. He was used to that.

About two weeks later, Robert and Ginger made their way down a path toward the nephrologist's office. A dialysis center was on their right. Through the long window spanning the side of the building, Robert could see a row of people sitting in big chairs. Some were asleep. Some were watching televisions mounted to the ceiling. Everybody looked old. Everybody looked miserable. Everybody sat next to a machine that they were hooked up to. The machines were big and loud.

His eyes stretched big and he turned to his mother and said, "If I gotta do that, you might as well kill me. I ain't doing that."

"Don't say that," Ginger said in a stern mother's tone, never wanting to imagine a time when her baby was not around. But then, softer, with hope in her voice, "Maybe you won't have to."

"Philipe said you had some protein in your urine," Barry Gorman said.

Robert nodded. He liked Gorman right away. He liked the casual way Gorman said "Philipe" instead of "Dr. Santoyo" and how instead of a white coat he wore his stethoscope draped around his neck like a horseshoe. He liked Gorman's bubbly personality. It matched his head full of lush salt-and-pepper hair. A clean-shaven face and tie were all he appeared to have in common with Santoyo.

"I want to find out what's going on," he added, and went on to ask a lot of questions and examine Robert.

"It could be nothing. It could be something," Gorman said after he finished examining Robert and after all his questions had been answered. He found nothing in particular that made him worry a zebra was afoot. With the exception of allergies to pet dander and pollen, Robert had been a normal, healthy kid. There was no blood in his urine, which along with the protein would definitely mean something. There was no pain or repeated bladder infections that would suggest he was born with abnormal kidneys. There was no swelling at his ankles. No one in his family was known to have a kidney problem. And for every ten kids with protein in their urine, it would be nothing to worry about in nine of them.

Nothing could mean that the urine was just concentrated, a deep, deep yellow, a sign that the kidneys are holding on to as much water as possible in a body that isn't drinking enough of it. The amount of protein in a very concentrated sample of urine would seem high, but once the concentration of the urine was accounted for, the actual protein content would be normal.

Nothing could also be a small amount of protein in the urine, trace or 1+ on a scale that goes to 4+, barely turning the color block on the dipstick from pale yellow to a light shade of green. After all, the interlacing fingers covering the glomerulus are not perfect—up to 150 milligrams of protein may slip through in a given day and still be normal. His twice-a-day football practices or a fever alone could have caused twice normal, enough protein leakage to turn the dipstick a deeper shade of green, 3+ green. But this would go away after a day of no strenuous exercise or when the temperature came down.

But a small amount of protein doesn't make your pee look like a freshly poured beer.

Still, even a persistently deep green 4+ dipstick, suggesting up to 2,000 milligrams of urine protein, could be nothing in a kid. It could be postural proteinuria, an exaggeration of the kidneys' normal tendency to let more protein slip through when the body is upright, which accounts for the majority of high urine protein on repeated dipstick tests in children. The first pee of the day, after lying asleep all night, would have normal protein. A full day of urine might have normal protein too. Nothing he wouldn't grow out of on his own.

But for every ten kids with protein repeatedly found in their urine, maybe one of them would have a zebra. Zebras with exotic, usually multisyllabic names. *Minimal change disease. Focal segmental glomerulosclerosis.* Still others joined the herd if you consider those caused by protein that sneaks past my imagined factory workers along the tubule conveyor belt. *Cystinosis. Tubulointerstitial nephritis.* Still others if you think about those that cause blood to leak into the urine too—sometimes too little for the human eye to see. *IgA nephropathy. Alport syndrome.* It was the one, the possibility of a zebra that obligated Gorman to investigate. A horse could not be assumed.

"To start we'll need to collect your urine for twenty-four hours," Gorman told Robert.

The protein was high in Robert's pee when he was still rubbing the sleep out of his eyes. The protein in his pee was high at noontime and nighttime. There were more than 3,000 milligrams in Robert's total pee for the day.

It was definitely something. Exactly which something it was could only be determined by looking at a small sample of his kidney under the microscope. Robert would need a kidney biopsy.

For centuries the only time kidney tissue was directly

examined was at autopsy. Gabriel Valentin's invention of the first crude microscope in 1837 allowed for descriptions of the kidney to include histology, features not visible to the naked eye. But the idea of obtaining tissue other than blood during life in the form of a biopsy was not described in medical literature until 1895, and then it was only within the context of skin diseases. The technique of taking samples of internal organs was honed with the liver biopsy, and it wasn't until the 1940s—when it was noticed that their attempts to biopsy the liver actually retrieved kidney tissue—that physicians began to wonder if the kidney could be deliberately biopsied using a similar technique. Swedish physician Nils Alwall attempted the first needle kidney biopsy in 1944, drawing upon the experience of Danish physicians Poul Iversen and Claus Brun, who pioneered the needle liver biopsy and were the first to publish on the technique of kidney biopsy in 1951. Alwall injected a special dye into the vein of the patient sitting upright and took an X-ray to give them a sense of where to insert their needle. This technique successfully retrieved enough kidney tissue to establish a diagnosis only about 40 percent of the time, which was not much better than an educated guess.

A portion of some of the poor success of these early biopsies was attributable to the needle itself, which aspirated or essentially sucked out the kidney cells, thus disrupting how parts were situated before biopsy. Robert Kark is credited with pushing the field forward in 1954 by repositioning the patient on his stomach and using a needle that produced longer, more intact samples. From there technology and technique advanced to our current use of automated biopsy needles and real-time ultrasound images, making percutaneous kidney biopsy safe and central to diagnosis of kidney disease.

But try to convince someone to agree to a kidney biopsy and watch eyes widen, brows furrow, forehead wrinkle, lips purse, and head cock to the side.

"It sounds worse than it is," I attempt to reassure them. "I wouldn't bring it up if I didn't think it would give us information that could change how I take care of you."

I go on to describe the procedure at our hospital in detail, hoping the more information I give them, the more the places in their mind that fear had filled with the stuff of nightmares might be replaced with the much less terrifying reality. I try to reach every possible nightmare scenario. Since it is hard to predict which fears each person has, and new fears seem to crop up all the time, I say everything to everyone.

I tell them that bleeding and infection are the biggest risks of the procedure and how we take special care to prevent them. Bleeding is more likely to happen if we don't go in the right place, so we position the person so that the kidney is easiest to get to. I tell them they'll lie facedown with a roll under the stomach so that the back will be flat and the lower part of the kidney comes down from under the ribs. And the radiologist will use the ultrasound—like what we use to look at the baby in a pregnant woman—to show us the kidney the entire time. It is rare that anyone bleeds so badly that we need to do anything about it or even give a blood transfusion. To prevent infection, we make sure there is no sign that the kidney is infected and that there is no skin infection where we plan to insert the needle. We clean the skin several times and lay sterile sheets around the area.

To reassure people that the pain will not be unbearable, I let them know we will numb the skin with lidocaine. To do that involves a tiny needle, so we will have to poke, and the lidocaine burns a bit. That lasts a few seconds. Then we will numb from the skin to the kidney with a very thin, long needle. We inject plenty of lidocaine as we go in. The kidney itself doesn't have nerves inside it, so they won't feel the biopsy needle inside the kidney. They will feel us pushing—we can't take that away—but we stop and give more lidocaine if there is even a hint of sharp pain.

I let them know that because the kidneys move up and down as we breathe, they will need to be awake in order to follow the radiologist's instructions on when to hold their breath for the few seconds that we are taking the sample.

Because many imagine we are removing big sections from each kidney, I let them know we will try to get two pieces from just one kidney. Each piece is about an inch long and about as thick as a pencil lead. To find out what is wrong, we need to get a sample of the kidney filters. A great sample is twenty filters. Each kidney has about a million filters.

I tell them that the procedure itself usually takes no more than a half hour, with the bulk of the time spent setting everything up. Afterward we have the person lie on their back for six hours—because the weight of one's own body will help prevent significant bleeding, and to monitor blood pressure and urine output. Finally we get a blood test for significant bleeding. If all is well, then we send them home to rest in bed until the next day and instruct them not to lift anything heavier than ten pounds for the next week or so.

In spite of all the detail and attempts to reassure, the occasional patient can't get past the fear of the possibilities. Or maybe not knowing what is happening in their kidneys is easier to cope with. They leave the clinic without making a follow-up appointment. Or they agree to the biopsy, we schedule and make all the arrangements—reserve the ultrasound suite, the radiologist, the recovery room bed—and then they don't show up.

When Robert had his biopsy done, computed tomography, aka the CAT scan, was used to visualize his kidney. He felt nervous when he went into the hospital that day. Not because he was afraid of needles—weekly shots for years had cured any needle phobia if one ever existed. Rather, it was the description he was given.

"It's like a controlled stab," Gorman had told him. This was not just another test. Perhaps it was the power of suggestion that made it feel exactly like he had been stabbed. Or perhaps it was that third attempt to get tissue that made Robert's lower half jerk that made it so.

"*Uh-oh,*" said Gorman when Robert jerked side to side. I imagine most if not all doctors have *thought* "*uh-oh*" at some point in their training or practice. But to not be able to hold back speaking it? Never a good sign.

"He nicked my colon!" Robert lamented. He had blood in his stool. The blood resolved in a day and Robert quickly forgave Gorman, but his fear of kidney biopsy would be permanent.

My experience in doing kidney biopsies has been very

different—at worst one patient had some nausea and bled enough to have temporary kidney injury but didn't need a blood transfusion. Every other patient said some variation of "It wasn't so bad" or "It wasn't as bad as I thought it would be."

"The biopsy showed a condition we call F-S-G-S . . . focal . . . segmental . . . glomerulosclerosis," Gorman said slowly to Robert and his mother in his clinic office about three weeks later.

"What the hell is that?" Robert asked.

Gorman went on to explain to him about how the glomerulus is the filter of the kidney, how the holes in his glomeruli were too big and that's why so much protein was leaking into his urine, and how some of his glomeruli were scarred down.

"What caused it?" Robert wanted to know next.

"We don't know. It could have been any number of things. An infection that was undertreated or maybe you just had it and it was never detected before."

Focal segmental glomerulosclerosis (FSGS) was first described in the 1970s. We have since learned that it accounts for 20 percent of heavy urine protein in children and 40 percent in adults. The vast majority of FSGS just *is*—a primary or idiopathic problem in how the interlacing fingers are made or work just springing up on its own (at least as far as we know). But it's estimated that about one of every five cases of FSGS is secondary—caused by a number of things such as specific viruses, drugs, other diseases, and about twenty gene mutations that are passed on from parents to children, like the two recently discovered variations in the gene that makes apolipoprotein L1 (*APOL1*). Exactly how they cause FSGS remains unclear, but how the mutations came to be is an example of Nature hedging her bets.

Normally *APOL1* is a substance in the blood that can destroy *Trypanosoma brucei brucei*, a species of parasite that is spread by the tsetse fly of sub-Saharan Africa and causes sleeping sickness. Over time, two subspecies of the parasite that were resistant to *APOL1*'s effect evolved. Nature responded with two *APOL1* variations that could destroy one of them. As a result, people whose gene carried the *APOL1* variant were more likely to survive the deadly sleeping sickness—and pass it on to their children. This is akin to the sickle cell trait, in which a slight variation in the gene that makes blood creates protection against the deadliest form of malaria, which is spread by the anopheles mosquito found in sub-Saharan Africa, South America, the Caribbean, Central America, Saudi Arabia, India, and Mediterranean countries. The downside of nature's response is that a child of parents who each carry one copy of the gene variant has a 25 percent chance of inheriting a copy from *each* parent. Two copies of the gene variant that protects against malaria is sickle cell disease, a form of severe anemia characterized by debilitating pain crises and a limited life expectancy, and two copies of an *APOL1* variant and the child is at higher risk of developing FSGS (4.25 percent chance over their lifetime) than someone with just one copy (.3 percent) or none (.2 percent). But from Nature's point of view, weighing the risk of having a child with sickle cell disease or FSGS against the risk of *dying* from malaria or sleeping sickness in days to months, the trade-off was well worth it. Researchers estimate that 51 percent of African Americans have at least one copy, but only 13 percent have two.

While it is academically interesting and really kind of cool that we can pinpoint which of thousands and thousands of genes can cause a particular disease, a burn sprays through the pit of my stomach, my heartbeat quickens, and my body stiffens every time I read or hear race-specific data.

When a gene is found to be more common in people of African ancestry, a series of assumptions and shortcuts takes place. African ancestry in the United States is renamed "African American," and is applied to everyone labeled as such, regardless of individual ancestry. Suddenly African American takes on a precision, as if it is the gene itself. As if African American can be precisely located in our DNA.

According to the Human Genome Project, which completed sequencing the entire human DNA in 2003, we humans are 99.9 percent identical. And the .1 percent difference does not fit in our defined race categories. Yet, amazingly, *in spite of all that science*, medicine can't seem to shake the deep-seated belief that there are real biological differences defined by our made-up race categories.

After this series of assumptions and shortcuts, one may arrive at the conclusion that an African American is an African American is an African American, when in truth African Americans include African, European, Native American, Asian, and Middle Eastern ancestries—and many within one body. This practice begs the same question that the medication BiDil, which was approved by the US Food and Drug

Administration in 2005 for the treatment of congestive heart failure *in Black patients*, did—how Black does one have to be for race-specific data to apply? Where exactly do we draw the line?

As late as the 1930s, states were adopting the "one drop rule" into law, which taught us that a person was Black if they had *any* sub-Saharan African ancestors because they embodied more than a single drop of "Black blood," thus meeting the "one drop rule." In modern US medicine, it appears that we employ a less scientific but more easily assessed version of the "brown paper bag test" (which was used by African Americans well into the twenty-first century to restrict elite organizations to light-skinned Blacks) to define "African American" instead: the "buff manila folder test," where Blacks lighter than the folder are assumed to be of White or Other/not-exactly-sure-how-to-label-you race.

This series of assumptions and shortcuts in medicine often restricts our thinking, making it lazy at best. We use race as a diagnostic tool—like a fasting blood sugar or an echocardiogram. A fasting blood sugar above 126 is diabetes. A part of the heart not moving like the others is a heart attack. A lot of protein in the urine of an African American is FSGS. Maybe, but Black people get other things too.

During my second year of nephrology fellowship, I remember telling the story of a young man with blood and protein in his urine before an audience of my peers and attending nephrologists that exemplified how race factors into our diagnostic reasoning. This was our "case conference," a weekly half-hour meeting in which each of us fellows were scheduled to lead a discussion of a case of unknown diagnosis to most of the audience. Medical school trains us to begin each case the same—a however-many-year-old Black/White/Asian/Hispanic girl/boy/man/woman with a past medical history significant for blah blah blah presented to the clinic/emergency room with this, that, or the other.

This day I purposely left out the race. *A twenty-six-year-old man with no significant past medical history presented to clinic with proteinuria and microscopic hematuria.*

I couldn't even get through the patient's recent medical history, which might give clues to why my patient had protein and tiny amounts of blood in his urine, before the most brilliant of fellows in my group raised his hand.

"What is his race?" he asked.

"Green," I said flippantly. "What difference does it make?"

My brief attempt at social commentary completely lost on him, he launched into a thoroughly race-based litany of what the diagnosis could be. "Well if he is Asian, IgA nephropathy would be most likely. If he is African American, then FSGS would be most likely. . . ."

*For real?* I thought. *Would knowing his race make you not get a biopsy?* I wanted to ask, but all other heads were nodding as if he were preaching what they believed. They too were all trained to believe knowing this patient's race was as important to establishing the diagnosis as knowing how much protein was in his urine or if it hurt when he peed.

My patient was White. I performed a biopsy. It showed IgA nephropathy.

An earlier experience during my residency stood out as an example of how the sword of race-based generalization could cut more than one way. The noon conference lecturer was a renowned rheumatologist. He presented a case of a forty-something-year-old Black woman with a cough, shortness of breath, weight loss, and large tender bumps on her shins. A plain X-ray showed enlarged lymph nodes in the center of the chest. All were classic signs and symptoms of sarcoidosis, an inflammatory disease that can affect many organs in the body—and is more common in Blacks than Whites and more common in women than men. Even we Highland Hospital bald-headed notch below redheaded stepchildren of UCSF could figure that one out. Yet, the renowned rheumatologist disclosed, the patient's diagnosis was delayed for years because he did not consider the diagnosis. He did not consider it because he was not really describing a forty-something-year-old Black woman. He—a sixty-something-year-old White man—was telling us his own medical history.

I don't know of a similar story in nephrology. But then again, we don't biopsy everybody—especially if we don't consider the possibility that the biopsy will show something different from the disease we have assumed.

"We're gonna keep an eye on it," Gorman went on. He explained that though Robert's glomeruli were damaged as evidenced by the protein in the urine, they were still working pretty well as evidenced by his creatinine blood test, used to estimate kidney function, which was still in the normal range—around 1.

"What does it mean?" Robert asked. "Am I gonna have

to be like those folks I saw by your office?" *If I gotta do that, you might as well kill me,* he thought again.

"Not immediately, but eventually."

"What's eventually?"

"Ten to fifteen years."

To a teenager, ten to fifteen years was forever away. He wasn't afraid because it wasn't immediate.

"How do I make sure it doesn't happen to me?" He looked at Ginger now. He could tell she was thinking the same.

Gorman handed Ginger prescriptions for a low-protein diet, an ACE inhibitor, and prednisone in hopes of lowering the amount of protein lost in Robert's urine. How useful a low-protein diet may be goes in and out of favor like fashion trends. ACE inhibitor medications lower blood pressure, especially within the kidneys to help lower protein loss.

Robert was able to tolerate the maximum dose of ACE inhibitor without the light-headedness from low blood pressure or the dry cough side effect that some people get with this class of medication, but he found the low-protein diet unsatisfying, and the prednisone gave him mood swings, a round face, and acne. The protein in Robert's urine didn't change very much up or down and his creatinine blood test stayed about the same after several months of treatment, so Gorman gradually tapered Robert off the prednisone and he was left with a chronic kidney problem.

Today we also try to treat the treatable causes of FSGS—recommending antiretroviral medications for FSGS caused by HIV, weight loss for FSGS caused by morbid obesity, stopping the heroin for FSGS caused by it, and a few fancier medications—but the end result remains the same. While some treatments work better for some people than others, eventually—unless the person dies of something else first—almost all cases of FSGS and the other zebras share a common path where cure is not possible.

They become plain old horses. Chronic kidney disease.

# 13

## HORSES

It is hard for most people to wrap their minds around the reality that what they have is *chronic*—as in it ain't never going away, never ever ever—kidney disease. They want me to prescribe the pill or shot that will make it go away. Tell them the food they should eat, the food they should stop eating to make it go away. Many look for a natural or non-Western approach to find their way out of the inevitable, assuming natural products are *only* safe and beneficial to health.

Traditional Chinese medicine is often painted with this brush. In this holistic approach, the kidneys represent not just the organs themselves but rather the entire urinary system and the hormone-producing endocrine system. The kidneys house the body's essence or life force (qi) and are the root of yin and yang for the entire body. While I appreciate the concepts of traditional Chinese medicine and am a big proponent of acupuncture for treating pain, for example, I am a bit leery of the herbal remedies central to it. I know it's been around for a couple thousand years, but I still don't know what's in most of it.

But what I do know is that plants containing aristolochic acid are commonly used in Chinese herbal medicine—and aristocholic acid can cause rapid scarring in the kidneys, often leading to end-stage kidney disease and cancers of the urinary system, making it one of the most dramatic examples of an herb that can damage the kidneys. And many other herbs can cause harm indirectly in those with or at risk for chronic kidney disease.

Some dietary supplements contain herbs that increase blood pressure or blood sugar, which could potentially cause or worsen chronic kidney disease. Others contain herbs that cause low blood sugar or high potassium, particularly in people with advanced chronic kidney disease. Still others contain herbs that can cause diarrhea and vomiting bad enough to cause severe dehydration, which can cause sudden kidney failure or acute kidney injury, which in turn places a person at higher risk of chronic kidney disease. But since people are less likely to think that any badness can come from natural remedies, they often don't tell their doctors they are taking them.

I'm feeling good, Doc," Mr. Holly said as I sat down on my stool in the clinic exam room. He sat in the chair beside the computer table, his round freckled face beaming.

I was not surprised. Most of my patients with chronic kidney disease tend to feel well until there is very little kidney function remaining because, like a woman losing herself because she is so focused on pleasing everyone else, She suffers in silence.

"That's great." I smiled back. But I had seen his lab results.

Mr. Holly was in his mid-forties and had fairly advanced chronic kidney disease as a result of diabetes. The diabetes had also been taking a toll on his eyes and heart, leaving him nearly blind in one eye and in need of a heart bypass surgery. He had ignored his health until five years ago when diabetes symptoms took center stage. But since I met him three years ago, he had been doing the best he knew how to improve his health—except to stop smoking or lose weight. At least he knew his medicines, was taking them, and showed up to his appointments. Well, most of them anyway.

I last saw him in clinic three months prior. His kidneys were filtering his blood at a rate of about 6 teaspoons per minute at that time, as it had been when I checked the visit before. Normal kidneys start out filtering about half a cup of blood per minute, or 25 teaspoons. At age forty, we lose about a teaspoon every five years (1 milliliter per minute every year) simply because parts don't last forever, but we don't really see evidence that She can't keep up with her responsibilities until the estimated function drops below 12 teaspoons (60 milliliters) per minute. So at 6 teaspoons Mr. Holly's kidney function was not good, but at least not worse than before.

But the labs drawn just a couple of days ago showed that his kidney function was down to almost 4 teaspoons (21 milliliters) per minute. If it dropped any further, I would have to start the dreaded, tear-inducing "We need to talk about what you want to do when your kidneys fail completely" conversation.

"What have you been up to?" I asked.

"Oh, I've been drinking this health drink from Costco," he said. "It's called açaí berry juice and it's natural and it's got all kinds of antioxidants and stuff in it."

Açaí berry. Açaí berry. I searched the recesses of my mind to remember that açaí berry can have the same effects as NSAIDs—nonsteroidal anti-inflammatory drugs—like ibuprofen, Motrin, and Aleve. Taking a lot of NSAIDs for a long period of time can cause sudden—and sometimes irreversible—worsening of kidney function, especially in people who already have chronic kidney disease. And though açaí berry can help patients with diabetes lower their blood sugar a little, lowering blood sugar while kidney function is getting worse can be extremely dangerous, because it is the kidneys that remove most diabetes medicines from our bodies. And because sick kidneys aren't so good at removing things, the medicines stick around longer—working to lower blood sugar.

"I've been drinking it every day for the last two months," he went on, sitting up straight with his shoulders back. He was pleased with himself.

"I want you to stop drinking it," I said sternly. His face changed with a quizzical arch of one eyebrow.

"You want me to stop drinking it?" he asked. I nodded and watched his frame deflate like a scolded little boy who knew he had been bad.

To ease his guilt, I explained that what he had done was not unusual. That most people assume natural products could only do good. That because dietary supplements are considered foods and not drugs, the companies making them are not required to test their products for safety in anybody, much less people with chronic kidney disease. They are not required to prove the benefits they claim the products have are true. They don't even have to make sure that each batch contains the same amount of an active ingredient.

"So, yes, please stop taking it," I said again. "And I want to check your kidney function again in one week." I hoped we weren't too late.

A week later his kidney function test was back to 6 teaspoons. I exhaled in relief.

I called Mr. Holly with the good news.

"So I shouldn't drink it anymore, huh, Doctor?" he asked, and I exhaled with something more like surrender before repeating myself again.

No one wants to hear that there is no quick fix for chronic kidney disease. What's best is to avoid it altogether. Don't gain too much weight, don't drink too much, don't eat too much salt, don't take too many NSAIDs, don't smoke or use recreational drugs at all, and avoid the majority of its causes including diabetes, high blood pressure, clogging of the kidneys' blood vessels, and HIV.

Some of the causes of chronic kidney disease are a bit harder or impossible to avoid, like cancer, kidney stones, or a large prostate blocking urine's path out of the body. Or being born with abnormal anatomy forcing urine the wrong way. Or developing a disease that damages the kidneys' filters like lupus or FSGS. Or inheriting polycystic kidney disease or sickle cell disease.

Once a person has chronic kidney disease, we try to slow down the rate at which the function declines by controlling or taking away the things that are hurting their kidneys. Control the blood pressure if it's high. Control the blood sugar if there is diabetes. Lower the protein in the urine with medications if there's too much in it. We've got lots of pills for all of that—three and four and sometimes five different kinds of pills for one body to take, once or twice a day, every day.

If we're lucky, I get to tell a patient that their kidney function is stable when they return for their next clinic visit, but there is potential for confusion in the word *stable*. A reminder came in the form of an e-mail.

"You said my mother's kidney function was stable and she didn't need to come back for two months, but someone called the other day and said her kidneys were really bad and that she'd need a kidney transplant," the e-mail accused. "So which is it?" it went on to demand.

"Both," I replied.

I know many people who became doctors because they wanted to "help people" and, truth be told, because they are fascinated by all the weird things that can go wrong with the human body. The teratoma, a tumor with hair, teeth, and sometimes even limbs growing in it, comes to mind. I don't know anyone who went into medicine to give bad news and upset people on a daily basis.

Yet this is where many of us end up—making people cry for a living. It kind of wears on the spirit after a while, which is why we tend to choose words that are technically the truth, hopeful even, but perhaps not very clear. Words like *stable*. *Stable* is a word doctors use in an attempt to reassure the patient that although this is an unfortunate situation, at least it hasn't gotten worse since the last time we looked at it. But at the same time it seems some patients interpret *stable* to mean things aren't too bad after all. Things are OK. Good even.

As chronic kidney disease worsens over time, we try to replace the functions kidneys stop being able to do. When there is anemia because the kidneys can't make enough of the hormone that stimulates the bone marrow to make blood, we can give a man-made version in shot form. When the kidneys can't maintain normal phosphorus levels in the blood, we've got pills to absorb it in food so it doesn't get absorbed into the bloodstream. When a patient can't buffer all the acid in the food we eat, we've got still more pills to balance it out. When they can't maintain normal potassium levels in the blood, we've got a powdered drink to help them poop it out.

But despite our best efforts, at some point, we can't keep up with the body's needs. The horses continue to trample until it gets to the point where She has suffered in silence too long and simply won't take it, can't take it, anymore. She gets sicker and sicker until, eventually, the person is left with end-stage kidney disease.

The hope is that our efforts will make *eventually* as far away as possible. Because once eventually comes, all that is left to do is to try to replace the kidneys with a transplant or dialysis, because we can't live without Her.

# PART IV

## YOU'RE GONNA MISS HER WHEN SHE'S GONE

# 14

## EXPECTATIONS

*He just needs a kidney and he's good, right?* I remember my friend Melva's words. Like Melva—and like most people—I thought transplant would be the cure for Robert's illness. I thought all our real problems would be over once Robert had a transplant. Transplant was supposed to be the beginning of our happily ever after. Our panacea.

But life and fellowship taught me that sadly, unfortunately, unfairly, with kidney failure there is no happily ever after. There is no cure. No panacea. I learned to see transplant for what it is—a treatment for kidney failure. And like any other treatment, it has its downsides. Bodies reject kidneys. Transplanted kidneys fail. The time in between is spent taking a handful of pills twice a day and silently praying that the disease that caused the original kidneys to fail in the first place doesn't come back.

One particular scare came after my nephrology fellowship, but it certainly wasn't the first. We were six years out from transplant, what Robert had come to call his re-birthday, when routine labs showed his creatinine had bumped. The blood test for kidney function was 2.5, when it had been less than 2 just a few months prior. Normal is closer to 1, but my kidney was considered small for Robert, so his kidney function had never reached normal. The bigger the body, the higher the muscle mass. More muscles produce more creatinine, so he needed more kidney to pee out the creatinine. A rising creatinine in the blood meant dropping kidney function.

Robert was dismayed. He had been doing everything he was supposed to do. How could this be happening so soon when we were hoping for forever?

There was little protein in the urine, so it was unlikely that FSGS was recurring in the new kidney. An ultrasound showed there was no evidence of a blockage like he had in that first week after transplant. Too much time had passed to worry about acute rejection—that usually happens within the first six months—but chronic rejection was still a possibility. Robert needed a biopsy.

Biopsy of the transplanted kidney is simple compared to biopsy of the kidney a person is born with. Located just beneath the muscles in the lower right belly, the transplanted kidney is much easier to get to with a biopsy needle, and it's much easier to apply pressure to prevent significant bleeding. To Robert, though, it was still a controlled stab. He grimaced his way through it and had to be reminded to breathe.

While he lay in the recovery room with a sandbag positioned on top of the kidney transplant, my colleagues invited me to join the team in the pathology department to look at the biopsy sample under the microscope. I was hoping they would—it isn't every day that a person gets to see their own kidney.

Looking through the microscope, I was transported back to the pathologist's lecture years earlier. I grinned. She was a thing of beauty. It was all I could do to not pop my collar. There was no sign of FSGS and no sign of rejection. I couldn't wait to give Robert the good news and almost skipped the whole way to recovery.

With obstruction, rejection, and disease recurrence off the table, the only thing left on the list of possibilities was high blood calcium, which was found on the routine lab tests. High calcium could cause blood vessels in the kidney to narrow, thus restricting blood flow to the kidney. To bring down Robert's calcium level required removing the source—overfunctioning parathyroid glands.

The parathyroid glands look like four slivers of rice nestled on the back of the thyroid gland, just above the breastbone. Because they work with the kidneys to maintain normal levels of calcium in the body, it is the norm for patients with advanced chronic kidney disease to develop what we call *secondary hyperthyroidism.* Here the parathyroid glands work harder, stealing calcium from the bones, as they are trying to keep up appearances of normal blood calcium in the face of failing kidneys—like Spanx fighting to maintain an hourglass figure against an expanding waistline. The bones weaken over time and are more prone to fracture—

unless the medications that try to replace the kidneys' role are taken diligently.

But some patients who have had end-stage kidney disease for many years go on to develop *tertiary hyperthyroidism,* in which one or more of the parathyroid glands have enlarged and overfunction to the point that calcium is abnormally high, in spite of medications—like a warped Stephen King version of Spanx that get thicker and thicker and can't be pulled off. At least two of Robert's parathyroid glands had made this transformation and could no longer be managed with medications. They needed to be removed as soon as possible.

Endocrine surgeon Qing Yu entered the clinic exam room with a relaxed purpose. There was no rushing, but there was no time to waste either. "I spoke with your nephrologist and reviewed your records," he said to Robert between his thin-lipped hello and sitting down. "I agree you need to have your parathyroid glands removed, a parathyroidectomy."

Though facing yet another surgery, Robert was not afraid. It was reassuring to know that Yu specialized in parathyroid surgery, but what abated his fear even more was the knowledge that the kidney was healthy. That what was wrong could be fixed. A surgery he could handle. But a failing kidney? He wasn't so sure he could handle that. Robert stared intently into Yu's round face as he explained what the surgery would entail. He liked that Yu gave him straight answers even when he didn't know.

"I will remove at least two of your glands. I may need to take more. I won't know until I get in there and see what's going on." He examined Robert's neck and showed him where he would need to cut. He warned him that if the glands were very low, he might have to cut his breastbone in order to get to them.

Truthfully, it was the cutting and the scar that would remain that worried Robert most. He already felt Frankensteinian with his fistula arm. He worried that he would have to walk around trying to hide his arm *and* his neck to avert the stares and questions.

"A plastic surgeon will be there to close the incision," said Yu, and Robert felt himself relax.

The surgery was scheduled within a few weeks. I expected it to last about four hours, so when the sixth hour came but Yu had not appeared, my anxiety surged. I had articles from medical journals with me to help pass the time, but all my eyes would focus on was the second hand ticking around the clock on the wall in the surgery waiting room.

Finally Yu emerged.

Robert's anatomy was complicated. Though tests before surgery suggested only two of his parathyroid glands were overfunctioning, once Yu exposed them, all looked enlarged. So a surgery that set out to remove two glands ended up removing three and a half. Yu hoped to leave just enough to keep Robert's calcium at acceptable levels. It didn't help matters that Robert's glands were unusually low, tucked underneath the breastbone. So low, Yu thought for a moment that he would need to saw his breastbone in half to get to them, but a wider cut and an aggressive pinning back of Robert's shoulders averted that threat.

Robert awoke to very sore shoulders and a five-inch-wide scar on his neck that would eventually fade to be barely noticeable, but also to the news that the surgery was successful. He was discharged the next day with a normal blood calcium level and a creatinine drifting back to where it was before.

"Where are you going?" I asked, watching him grimace as he began to pull himself up from the couch at home.

"I'm going to get something to drink," he grunted.

"You know . . . I can do that for you. Sit back," I said as I rose and headed to the kitchen.

"Oh," he said, his face scrunching either because of pain or because it hadn't occurred to him that I wanted to take care of him, but he did sit back. It was hard for Robert to let me dote on him. He wasn't used to people doing anything for him and had learned early in life not to expect it.

"You're a self-starter," his parents had said to him when he was a boy. "Your brother needs extra help," they would say, as they proceeded to rescue his two-years-older brother Eddie from predicament after predicament. Little Robbie was expected to fend for himself. Eddie was presented a new car when he turned sixteen. Robert bought his first car on his own after college. Eddie's four years at community college, then five more at Morehouse, were parent-financed. Robert took loans to cover what his football scholarship would not. As a result, he prided himself on not needing anyone, on not being indebted to anyone.

A failing body took away that pride. Every need that kidney failure created put another crack in the mirror of how he saw himself. The reflection had become so distorted, it was hard to remember what he once looked like. Who he used to be. He had *needed* to live with his parents in order to feel safe. He had *needed* a dialysis technician to stick his fistula just right in order to have just an OK day. He *needed* dialysis or a kidney transplant in order to live.

Some needs were easier to accept. Parents are expected to help their children. Dialysis technicians are paid to do a job. But nobody has to give away her kidney. It is a gift that cannot be matched—which was why Robert would have preferred a deceased donor kidney. A dead man collects no debts. My kidney made Robert whole again, gave him hope that he would get back to who he once was and on to the person he believed he was supposed to be—but shards of mirror fell to the floor. Robert felt indebted.

And every April 14 since 2005 was a reminder of it.

I had two points of reference for what our transplant anniversary could look like. The first was through Amy Markowitz, a scientific editor with a law degree whom I had been introduced to when I was just planning to donate my kidney. She described how years prior she had donated to her best friend. And how every year since then, her best friend had spent the day expressing her gratitude in words, time, and togetherness. Gratitude that because of what Amy did, she never had to go on dialysis. That because of what Amy did, she would have a much better shot of watching her two young children grow up.

The second was during my nephrology fellowship. The recipient lay in a gurney as I helped prepare him for a kidney biopsy. It was the third or fourth biopsy he would have in as many years since his wife gave him one of her kidneys. Another rejection scare.

He grinned and shrugged his shoulders, *Guilty as charged,* like an irresponsible teenager as she stood at his side lamenting the fact that she didn't know what else she could do to get him to take his antirejection medications consistently. I had shared that I donated my kidney around the same time she did.

"Sorry, I don't have to remind my husband to take his medications." I shrugged, then added, "I did my part in giving him a kidney."

She looked at me as if it hadn't occurred to her that she had done her part too. That remembering to take the medications to keep her gift to him alive was the *least* he could do in return. She turned to look at him as if to say *See, her husband remembers to take his medicine.* He looked back at her with the same stupid grin on his face. He was literally pissing away her kidney, as if she had another to give. I couldn't think of a more disrespectful disregard for what she had done.

As Robert and I approached our seventh transplant anniversary, Robert's shrug in response to my question of how we would celebrate it felt painfully close to the man pissing away his gift. I took his shrug to mean he hadn't given any thought as to what we would do, as if it was no big deal to him. He intended his shrug to question why I was asking him for the plan, as if it was his responsibility alone. We both reacted. And argued. About how we would celebrate the anniversary of what was arguably the most pivotal day of both our lives.

Truth is, neither of us can remember what was actually

said that day. Only how we interpreted the words. Only how they made us feel.

He heard me say that the transplant anniversary was all about me, when he felt it was supposed to be about us and what we had been through together. He heard me telling him that he *had* to do something for me like *I* wanted it done, when he felt taking care of the gift was a daily demonstration of his gratitude. Life had taught him that only actions spoke the truth. Words were just background noise.

I heard him calling me selfish for expecting a simple thank-you once a year, when I felt it was the *least* he should *want* to do. Life had taught me people do things for many reasons—because they have to, because they are expected to, because it's good for them to. It was the words that bubbled up directly from your heart and got stuck in your throat that held the real truth.

I heard him say that if I only did it because I felt like it obligated him to say thank you every year, then I shouldn't have done it. Because he never asked me to. I heard him say this like he had been holding the words back for six years.

# 15

## READY OR NOT

Robert's transformation from choosing death over dialysis to choosing dialysis over death when death was imminent was not unusual. I have learned that this is the modus operandi of my patients with worsening chronic kidney disease much more often than not. Everybody thinks that somehow they will be the exception to the outcome the doctors are promising. That somehow it will go differently for them.

I believe we all do this in some form or fashion. The teenager who believes he can try meth one time and his life won't fall apart as did the lives of everyone else who went down that path. The man who believes he will beat the pancreatic cancer he was just diagnosed with when everyone else who had it before him died within a couple of years.

I've done it too. I remember watching Greg Behrendt on *Oprah* and then the movie version of his book *He's Just Not That Into You* with great interest. As a woman, the message struck a chord in me because I had made many, if not most, if not all the mistakes Behrendt points out that most women tend to make in search of "the one." I had been guilty of reading way too much into any little thing to convince myself I was on the verge of something really beautiful from boyfriend #2 through husband #1, when truth be told, they simply were not that into me. I was not the exception to the rule that unnecessarily complicated and ill-defined relationships never turn into something beautiful. I was the rule.

I see the same misguided exception thinking in my encounters with patients I am encouraging to begin preparing for the inevitable, impending need to start dialysis. It avoids ER trips, hospital stays, avoidable procedures. All too often these conversations end with their rejecting the notion out of hand. Most often their rationale for doing so is a firm belief that [insert higher power's name here] will prove me wrong. They find meaning in imagined things—*My legs seem less swollen today, so maybe that is a sign my kidneys will get better.* They downplay the symptoms they are having—*I'm not that nauseous.* They hope and pray He will not allow them to be just like every other patient in the exact same situation. That they are somehow different from the rest. That He has a plan for them that will make them the exception, when I want to tell them the unfortunate truth that "*He's* just not that into you. You are the rule." But what I actually say is "I wish it could be different" or "I know this is hard" because this is the truth too.

I saw Talia Afuta in our pre-dialysis clinic for patients with advanced chronic kidney disease. She was nearly a year from her fiftieth birthday, but her brown Samoan skin was smooth like her hair, which was coiffed just so into a low pompadour slicked into place with gel. I walked into the clinic exam room to find her face clenched like she was silently counting down the seconds until she would get up and leave. I apologized for keeping her waiting so long.

She felt fine, she said. No problems.

I was glad to hear she felt well. But by the look of Ms. Afuta's blood test results, I was skeptical that she felt as well as she claimed she did. Either she had just become used to feeling bad or she didn't want to own up to how bad she did feel.

It ain't never good for one's hemoglobin (anemia test) to be the same as one's kidney function test. Usually people with a hemoglobin of 7 (normal in a person without kidney disease is closer to 13) feel tired or short of breath or have chest pain when exerting themselves, but she said she did not. She was a large person. Usually when a large person's kidney filtering rate drops to less than 1½ teaspoons (7 milliliters) per minute like hers had, they felt nauseated, sometimes vomited, had a metallic taste in their mouth, and felt exhausted. A smaller-framed person might still feel well at this level of kidney function because smaller bodies can get by with less kidney. There is less body to remove wastes from.

But maybe this large-framed "not Black" woman's kidney function was closer to the race-adjusted estimate of almost 2 teaspoons (9 milliliters) per minute and she really didn't have any symptoms yet. She only admitted to *maybe* a slight metallic taste—but that was caused by the pills doctors told her to take, she said. She was sure of it. Doughy flesh spilled over her feet without a hint of an ankle—but that was only because she was on her feet a lot, she said.

I explained her lab results. Gave my advice to prepare for dialysis, which was long overdue.

"I told my primary care doctor and the other doctor here—I'll tell you when I am ready," she said, averting her eyes.

I took in a deep breath and exhaled slowly. *Here we go again*, I thought. Yet another patient with advanced chronic kidney disease refusing to do what was in her best interest. Yet another patient refusing to take action to prevent a bad situation from getting worse. As if a miracle could still happen.

Had Ms. Afuta been facing end-stage kidney disease just sixty years ago, before dialysis existed, she would be dead within a few months at best. The miracle was dialysis because She ain't so easy to replace. A miracle that was centuries in the making.

It wasn't until the sixteenth century that scientists began to notice that when urine could not pass, as in people with stones blocking the urinary system, the blood became a fluid more like urine. Over the next couple hundred years, they realized that the urine contained wastes that, if retained in the blood, would lead to the sickness and eventual death of the person with kidney failure. It wasn't until the eighteenth century that they noted a "soapy" substance in the blood that they called "urée" and thought was the waste product responsible for what they called "uremia"—a condition that followed kidney failure in which the chemicals in the body were out of balance. It was another fifty years before they realized that, even though large amounts of this substance called "urea" were in the blood of those with kidney failure which increased progressively until death, injecting it into the blood of animals produced no toxicity. It just made them pee more.

In the 1830s and 1840s, French scientist René Dutrochet discovered the natural movement of water from areas of lower concentration of substances to areas of higher concentrations—while membranes prevented the passage of the substances. He called this *osmosis*. This work earned him the title of intellectual grandfather of dialysis, as it marked the beginning of the thinking that wastes might be removed from the blood by this process. It was Scottish chemist Thomas Graham who is credited as the father of *clinical* dialysis because in the 1850s and 1860s he took Dutrochet's work further by describing the separation of substances across membranes, renaming it *dialysis*, and observing that urea could be dialyzed across a somewhat porous membrane. This set the stage for future scientists to consider dialysis specifically for the treatment of uremia.

Still, it took several more decades of sorting out how to make membranes and how to keep the blood from clotting once it was outside the body. It wasn't until World War II that Willem Kolff—the father of dialysis—developed the first artificial kidney in the Netherlands in 1944. It was a huge wooden rotating drum with six feet of dialysis tubing wrapped around it that sat in a tub. Wooden because metal was taken over for the Nazi war effort and rotating so that blood could flow through the tubing. But getting steady access to the patient's blood meant surgeons had to *cut down* from the skin to the pulsating radial artery in the wrist—not something that could be done over and over again for weeks or months on end. So dialysis was only feasible for patients with reversible kidney failure with the patient just needing a few days of support while their kidneys recovered from a temporary damage.

But in 1960, Belding Scribner and Wayne Quinton in Seattle had a breakthrough that led the way to making long-term hemodialysis treatment for patients with irreversible kidney failure possible by developing the first working shunt. Made out of the new at the time nonstick material called Teflon, this first arteriovenous connection was a tube sewn into an artery and another into a vein connected in a loop by another piece of tube so that it lay on the skin between dialysis treatments. However, it lasted weeks at most. It wasn't until 1966 that James Cimino developed the technique still used today for the first surgically created arteriovenous fistula, making it the first hemodialysis access that could last for years.

In the meantime, other scientists were trying to avoid the blood altogether by developing a means of doing dialysis using the body's own membranes. While the intestines, the bladder, the space between the lungs and the skin, and even the spinal fluid were tried, the peritoneal membrane lining the abdominal organs proved to be the best. Jacob Fine led the most important work in peritoneal dialysis at the department of surgery of the Beth Israel Hospital in Boston. The first patient to recover from reversible kidney failure using peritoneal dialysis was in 1945. Their original system instilled fluid similar to plasma, the liquid part of blood, into the abdominal cavity between peritoneum and skin with one tube, while the fluid drained from another—continuously for four days. It took another four years to figure out how to do it with just one tube being poked through a patient's skin.

We've come a long way from these first efforts of hemodialysis and peritoneal dialysis, even though to this day we don't fully understand the exact nature and scope of uremia. Nevertheless, this work made the kidney the first organ for which *complete* mechanical replacement was possible. It took about four centuries.

Yes, dialysis was the miracle.

No doubt Ms. Afuta had a different kind of miracle in mind, but dialysis was the only one that was coming. Without it her sixteen-year-old baby girl would soon be motherless. She seemed to be looking at dialysis as if it was the end of her life, when in truth dialysis would just mean her life would be different. No, it would not be perfect. No, it would not take away all the issues that come with a vital organ dying. Yes, it would take time and have its frustrations to deal with. But the time in between could be filled with life. Family. Friends. Exercise. Work. Play. Travel. Joy. Love. With dialysis she could expect to live years, even a decade or two. With dialysis she could expect to live until a kidney transplant was available. It would take months to arrange if she had a living donor, years if she didn't. Without dialysis she didn't have months, much less years. There was no escaping dialysis. If she wanted to survive. If she wanted life.

I wondered if there was something in particular she was afraid that dialysis would do to her. I have been surprised to learn some of the misconceptions patients have about dialysis. The misconceptions are the fears, and the fears can be paralyzing. They prevent the patient from taking the well-advised steps toward preparing for dialysis. Paralysis creates emergencies.

*"They said dialysis 'cleans the blood.' I think dialysis burns the blood to clean it"* and *"They said your blood goes into the machine to take out extra water and toxins and then brings it back to your body. I am afraid I will die while all my blood is taken out of my body"* are two of the more extreme ones I've heard.

So I've learned to give more information to preemptively allay some fears. "Dialysis removes wastes and extra fluid from the blood by passing it through a filter. The filter is an artificial kidney," I'll say. "No more than two cups of your blood is outside of your body at a time."

Since I can't be sure that I've reached the person in front of me, I add, "Sometimes people think things about dialysis that frighten them but simply aren't true. If you tell me what frightens you about dialysis, maybe I can help put your mind at ease."

Ms. Afuta said she understood what dialysis was and had no questions about it. She knew she couldn't escape dialysis *and* live. She could even articulate that waiting would mean the emergency room and rushing to things that could be avoided with controlled preparation. Yet she could not reconcile these knowns with an overriding inability to voice acceptance of something she wanted no part of.

I tried to see the situation from her perspective. I acknowledged how hard it must be to come here and listen to bad news followed by more bad news from a person she didn't know from Eve.

"I wish I had known you for the last year or two, so we could have had time to get to know each other, so you could know that I am telling you what is right for you," I said. But what I really wished was that she didn't have to be in that situation at all. At least not so soon. That she had been screened for kidney disease early on when she was just "at risk" with a history of kidney disease or diabetes or high blood pressure in her family. Then, simple blood and urine tests once a year would have been all it took to find out if a hint of kidney damage was at hand. Then our efforts to slow things down would have been more like sitting on a hill with a beach ball against our back, our task to keep it from rolling down. Instead, we were left with a boulder, pressing against our back, gaining weight and speed as it forced us down the hill despite heels dug in.

Her brown eyes looked into mine as if maybe she could trust me. Soon.

Soon hadn't arrived by the time I saw her again three weeks later. Still she said she felt fine, that she wasn't peeing less or swelling more, though I could hear the crackling sounds of fluid beginning to collect in her left lung. But still she wasn't ready. Maybe next visit, she said. I sighed, disappointed. It's hard to watch someone drowning, especially when you've extended a life buoy. I told her what symptoms to watch for that would mean she should go to the ER before I saw her again.

Another three weeks later, on my office computer screen, I saw her name on the schedule for clinic that day and wondered how she had been doing since the last time I saw her. Though I hoped her kidney function was stable and that she was still feeling relatively OK, I hoped even more that she would be willing to allow me to make the arrangements to get her started on dialysis without having to be admitted to the hospital. Didn't matter to me if I was preparing her for peritoneal dialysis or hemodialysis. The quality of dialysis and the life expectancy of peritoneal dialysis are the same as they are with hemodialysis. It boils down to personal preference and a few medical considerations, such as morbid obesity or prior surgeries on the belly that make peritoneal dialysis ineffective. Whether the person preferred to be at home doing their dialysis either themselves or with the help of a caregiver as it best fit into their schedule or would rather come to a nurse- and technician-staffed center at a scheduled time for hemodialysis.

In a 2012 study, US nephrologists were asked which modality or way of doing dialysis they would choose if they themselves were faced with kidney failure and a kidney transplant was five years away. Only 7 percent said they would choose in-center hemodialysis. Yet more than 90 percent of dialysis patients in the United States receive in-center hemodialysis three times a week. This has more to do with an American expectation of being taken care of rather than taking care of one's self, the lack of a payment structure for nurses to help those who can't physically do their own dialysis at home, and what the nephrologist is most familiar with than anything about peritoneal dialysis itself. People balk at the fact that home dialysis must be done every day, whereas in-center hemodialysis is just three times a week. If one considers the total time required of each type of dialysis (for example, including travel to and from in-center hemodialysis), then peritoneal dialysis can actually take less time than in-center hemodialysis. But time isn't really the issue. Most patients presented with the options choose to accept the needles and symptoms of fatigue, dizziness, and cramping that come with trying to do in just three or four hours (as with in-center hemodialysis) the work that peritoneal dialysis does all day—without needles or these symptoms—in exchange for only having to think about dialysis three days a week.

In other countries, 25 percent to more than 70 percent of patients with end-stage kidney disease do peritoneal dialysis because hemodialysis is too expensive. In America, peritoneal dialysis costs about $67,000 per person per year and hemodialysis is about $85,000 per person per year. Here we seem to assume cheaper is worse when really it just means there is no staff or building that must be paid for.

Any dialysis method Ms. Afuta had chosen would have been fine by me. Already it was too late to avoid a hemodialysis catheter, even if she had surgery to create a fistula the next day. The reality was that it would take weeks to get her an appointment to see the surgeon and have the study of her blood vessels to determine which were large enough to make a fistula, more weeks to actually have the surgery, and at least six more weeks before the fistula would be ready to use. We may not have been able to avoid the hemodialysis catheter even if she wanted to do peritoneal dialysis, as that required an appointment with the general surgeon or interventional radiologist and two weeks to heal before being ready to use.

On my way to clinic, I stopped by the hemodialysis unit to write an order in one of my patient's charts and was surprised to see TALIA AFUTA in block letters on the spine of a neighboring chart. Her name was hand-written—the clerk had yet to find the time to print it out from the computer onto paper color-coded to signify who her nephrologist was. She had been assigned to the Tuesday-Thursday-Saturday third shift. She was due to come in for dialysis later that day.

After clinic I stopped back by the dialysis unit to find out how she went from "You should get ready for dialysis" in the clinic to "If you don't start dialysis now you will be dead within hours" in the ER.

She sat in the recliner with the leg rest extended. Her skin was ashen. She appeared to have aged a lot since I saw her not even a month prior. She was watching a movie on a DVD player. She looked up as I approached, and the look on her face changed from contentment to embarrassment, like she knew a well-deserved *I told you so* was coming her way. Instead, I asked her how she was doing and what happened.

"I tried to lay back in bed and I felt like someone was trying to kill me," she said. She was describing a sensation of drowning just like others with fluid building up in their lungs described.

"I told my boyfriend to call 911 and he freaked out," she went on. "And can you believe they kept me in the ER for more than twelve hours before I got a room?" Her tone was higher pitched with a twinge of anger then.

"Yes, I can." I nodded knowingly. "I was trying to spare you from that," I said, not being able to restrain my *I told you so* after all.

Just as Robert wasn't unusual in saying he'd rather die than start dialysis, Afuta wasn't unusual in resisting advice to get ready for it. Many patients hold on to the notion that they feel just fine and they pee just fine and they would rather die than *ever* start dialysis, despite the assertions of our clinical expertise to the contrary. They believe it is the dialysis that will kill them, not the end-stage kidney disease. They saw it happen to their mother/friend/second cousin on their father's side. The dialysis done it. Not Mama's heart attack. Not cousin's stroke. And definitely not drug buddy's continued cocaine, heroin, or methamphetamine use.

But when the kidneys *do* fail, a strange thing happens. Suddenly they don't feel fine anymore. It becomes too much to deny. It's at this point that fear of death overrides fear of dialysis. Dialysis, they come to see, is the thing that will save them.

# 16

## CHANGING TIMES

In 1960, the first dialysis center for patients with irreversible kidney failure—end-stage kidney disease—was established in Swedish Hospital in Seattle. It had the capacity to treat just five patients at the time, when more than one hundred thousand people were dying from end-stage kidney disease each year. Dialysis was a limited and expensive ($10,000 a year in the beginning, the equivalent of more than $80,000 today) resource, so someone had to decide who would be lucky enough to get it. That someone became known as the God Committee.

The God Committee—or the Admissions and Policies Committee of the Seattle Artificial Kidney Center at Swedish Hospital as it was formally named—was a seven-member group appointed in 1961 by Seattle's King County Medical Society. A banker, a housewife, a labor leader, a lawyer, a minister, a state government official, and a surgeon were tasked with deciding the select few adults who would get this new lifesaving treatment, which lives were worth saving.

By 1967, there were roughly one hundred dialysis centers around the country able to care for at least one maintenance dialysis patient. But there still wasn't enough capacity to take on everyone in need. A set of medical and social worth criteria had to be met in order to get dialysis.

One would not have to understand much of US history to guess who was deemed worthy. Young. White. Male. These were the characteristics of the vast majority of those first few thousand maintenance dialysis patients. According to 1967 dialysis census data published in the *Journal of the American Medical Association* in 1981, 91 percent were under age fifty-five, 91 percent were White, 75 percent were men. But when the magnitude of end-stage kidney disease in the population meant that even some of the worthiest were dying, legislation in 1972 was passed to extend Medicare benefits to provide dialysis treatment to essentially every citizen with a nephrologist's order—the Medicare ESRD (End-Stage Renal Disease) Program. By 1978, the dialysis population was much older, of color, and female. Nearly half were over the age of fifty-five. Blacks made up a third of dialysis patients, consistent with the fact that Blacks made up a third of the population that had end-stage kidney disease. There was no longer a gender disparity. This new law promised equity in dialysis delivery, but also led to a system where it could be a government-funded option for any US citizen with kidney failure—even Spider-Man.

Spider-Man is the name that came to mind when I saw my dialysis patient Jerry Shaw standing for the first time. For the most part, I only saw my patients in recliners as they received their dialysis treatments, not walking around or in a clinic exam room. So when I walked into the dialysis unit while Jerry stood waiting for the technician to finish cleaning his chair after the last patient, I was taken aback. He looked like he would topple over at any moment, clearly and heavily under the influence of Valium or methadone or heroin or something else or some combination thereof. His limbs and head seemed to hang from an invisible thread attached at his upper back. His belly bulged beneath a bright yellow Hawaiian shirt. He looked like a spider.

Jerry had bipolar disorder and a heroin addiction but refused psychiatric care and drug rehab programs. The only thing that helped, according to him, was double doses of Valium. He said they eased his anxiety, but from my perspective they maintained the semiconscious state he preferred to exist in.

I wasn't always so cynical about Jerry. In the beginning, I tried to reach him. He was the first dialysis patient I met in my new role as nephrology faculty and I was caught up in that same thinking that many doctors can be guilty of—that no one before me had really tried to reach him and if they did they didn't do it as well as I could. I don't know if this attitude comes from youth or cockiness—probably a combination of both. I pulled up a medical stool and gave him time to show him I cared. I even reached into his filthy bag of pill bottles and got my finger smeared with something of the color and consistency of peanut butter. At least I hoped it was peanut butter. Some bottles weren't capped and many were full when they should have been taken months prior. I visited him when he ended up in the hospital for an infection in the blood. "I didn't shoot up *into* my fistula," he admitted when I was trying to determine the source of his infection. "Just *near* it."

Perhaps it was this long hospital stay that began with the discovery of pus coming from his fistula that led to an infection of his spine and heart, followed in less than a week by another hospital admission prompted by his crack smoking–induced lung failure that caused my bleeding heart to run dry. I began to see him as unreachable, offering nothing positive to anyone around him or even himself. He actively, *repeatedly* engaged in behavior that he knew could take his life—which seemed a lot like suicidal behavior to me. But because he would be found unconscious and brought to the hospital, dozens of health-care providers and hundreds of thousands of dollars were dedicated to saving him. Repeatedly. Eventually he was found dead in his SRO (single-room occupancy) hotel room.

The Medicare ESRD program entitled him to receive in-center hemodialysis at the modern times rate of nearly $85,000 a year. All with nothing required in return—not even that he engage in counseling or rehab or anything that might help prevent another attempt to take his own life.

There was always a Jerry or two under my care. But even though the Jerrys felt like a substantial part of my practice because of the disproportionate share of time and money we gave them, only about seven thousand of the more than one hundred thousand patients starting dialysis each year have a drug addiction.

However, because both transplant and peritoneal dialysis are considerably less expensive than hemodialysis, the Jerrys demand the most expensive care by default. I have yet to see a patient with end-stage kidney disease and an active drug addiction be managed with peritoneal dialysis. Being able to do peritoneal dialysis at home first requires that a person have a home and one with enough space to store dialysis supplies. Homelessness is common among patients with drug addiction, and an SRO hotel room can't accommodate the twenty-plus large boxes of dialysis fluid that are delivered each month. And since ongoing illegal drug use can cause kidney disease and those using illegal drugs tend not to be reliable about taking medications, kidney transplant is a nonstarter.

Moreover, injection drug use damages blood vessels, usually making creating a fistula impossible. Just by virtue of the catheter being a foreign object piercing the skin, it is prone to infection. In the Jerrys, a catheter also presents easy access to the bloodstream—just a couple of twists to get a tiny cap off when he would have otherwise had to go hunting between his toes for a vein to get that next high. This greatly increases the risk of overdose and infection and long hospital stays. Therefore the Jerrys also demand care that places them at even greater risk of complications and death expected from end-stage kidney disease alone.

The Institute of Medicine's *Kidney Failure and the Federal Government* states that cost containment should not be a criterion for deciding whether or not to offer dialysis because dialysis can be lifesaving and is readily available. What began as one dialysis center able to provide treatment for five patients had grown to 5,800 dialysis centers capable of providing more than 43 million dialysis treatments each year by 2012. As the number of people with end-stage kidney disease requiring dialysis rises—as it does every year—so has the number of dialysis spaces grown to keep pace. That's nearly five hundred thousand dialysis spaces. No longer is there a question of whether any citizen deserves dialysis. The question begins and ends with "Does the person have kidney failure?" If yes, then yes, dialysis can happen.

Patients with end-stage kidney disease make up less than 1 percent of the Medicare population but consume about 7 percent of the Medicare budget. So I can't help but wonder, what price—in human effort and dollars—is too much to pay to maintain a life that is actively, repeatedly trying to destroy itself?

I believe the least we could do is require that patients with drug addictions engage in drug treatment. Addiction specialists maintain that drug rehabilitation programs are readily accessible and effective for the treatment of drug addiction. What percent of dialysis patients with drug addiction participate in drug rehab programs is not known, but in a 2013 survey from the National Institute on Drug Abuse, less than 15 percent of people with drug addiction in the general population received treatment in the past year. What's worse, 80 percent didn't believe they needed treatment.

There are some situations in which patients are expected to demonstrate that they are committed to successful outcomes in their own care. Many transplant centers require obese patients to lose weight and patients with drug or alcohol addiction to show they can get and stay clean and sober before they can be considered appropriate transplant candidates. Morbidly obese patients are required to lose significant weight prior to bariatric surgery to demonstrate commitment and improve continued success after surgery. It would seem logical to require patients with end-stage kidney disease and drug addiction to actively engage in drug rehab programs as a part of maintenance dialysis care for the individual's good and for society's good.

I question if this will ever happen because here in America we believe in individual choice, no matter the cost. And, contrary to the volumes spoken by their actions, the Jerrys are not suicidal. They are afraid to die. Individual desire trumps societal good.

But, as I was taught when I was a fellow, there were still a few situations in which we did not offer dialysis. When it simply was not appropriate. Cases where dialysis was considered futile because another devastating, untreatable, life-limiting illness was at hand.

I stared at Mr. Madani's CT scan images on the monitor. The horizontal slices through Mr. Madani's fifty-five-year-old torso showed that the cancer that started in his left kidney had taken over most of his abdominal and pelvic organs. Right kidney. Pancreas. Liver. The sheer bulk of the cancer had forced the diaphragm separating abdomen from chest upward, shrinking the space that his lungs could expand into.

The surgeon's attempt to cut out the cancer when it was thought to be just in the left kidney had clearly failed. There was nothing else the oncologists could offer him. Nothing they had tried—and they had tried everything—had stopped the cancer from growing. The last chemotherapy they tried only succeeded in poisoning his remaining kidney. They knew this could happen, but I imagine they hoped his right kidney could withstand the assault while the cancer alone retreated. The normal kidney function test just a week prior suggested this hope in a story of dashed hopes was reasonable. Wrong again. The kidney was failing, so the oncologists sent him to be admitted to the hospital for nephrology consultation.

At this time I was the nephrology fellow, responsible for gathering the pertinent facts and physical examination data to report back to my supervisor, the attending nephrologist. *This will be an easy one*, I thought as I left our workroom and headed for the patient's. *This guy is dying. I don't even know why they called us for a consult. They know we don't dialyze people with terminal cancer.*

After three obligatory taps on the closed door and a brief pause to announce my presence, I walked into the hospital room, sure I would be leaving very soon to be on to the next task in another long, busy day. This would be a quick no to dialysis, and I wouldn't even need to see him tomorrow—a rare one-visit consult.

The private hospital room, standard at UCSF Hospital,

felt large around the emaciated Mr. Madani. He sat leaning slightly forward in a blue vinyl waiting room chair while the recliner stood empty. His elbows were propped on the black metal armrests. His shoulders hunched in his singular effort. Breath in. Breath out. The pale blue hospital pants lay baggy over bony legs. He looked up at me without moving his head, eyes sunk deep into his skull. The whites were a tired yellow.

I introduced myself and my purpose for being there. He nodded ever so slightly as if movement would only make things worse. By the window stood Mrs. Madani. Her knitted brows untwisted and lifted with the hope of what my presence might bring. I nodded a thin-lipped hello.

"How do you feel?" I asked, turning my attention back to Mr. Madani.

"I itch," he replied, looking up at me momentarily, then back to staring at the white linoleum tile a few feet in front of him. The itching seemed to consume him, demanding his constant attention. He did not bother to scratch. I imagined the itch bored deeply into places his nails could not reach.

Itching could be caused by kidney failure, but according to his chart the itching predated the kidney failure by weeks. In Mr. Madani's case, the itching was caused by a buildup of bilirubin, caused by his cancer pressing the draining bile duct shut like a pinched straw. Several attempts to pass a tiny expandable coil across the pinch to stent the duct open had failed.

I walked over to him for a basic examination, trying not to make him move. None of the usual thumping, probing, or prodding in a search for clues to solve the medical mystery. I already knew the past, present, and future.

I stood at the back right corner of the chair between him and the made bed. A bony spine peered through the faded blue hospital gown ties at his nape and lower back. My stethoscope chest piece at his midback. Decreased breath sounds. No crackling sounds of fluid in the lungs. My stethoscope just right of his left nipple. A heart beating with a regular rate and rhythm. My right hand pausing for just a moment on his firm, bulging abdomen. My right thumb releasing its pressure against his right shin. No dimple of displaced fluid left behind. Done.

"OK," I said as I stepped away. "I have to go talk with my attending to decide what to do." I tried to appear as neutral as possible.

Mrs. Madani, still standing at the window, turned to face me fully, her hands clenched together. Tears in her eyes. "Anything you can do, Doctor," she pleaded. Mr. Madani blinked in agreement.

The love and desperation in her words and eyes took my breath like a basketball hard to my chest. The prickly sting of my own emotion crept up into my throat. I could not speak. I nodded a thin-lipped *We will do the best we can*, knowing that by *anything*, she wanted us to do something that could maybe extend his life by any means necessary. I almost wanted to grant her wish.

I wondered how much of the situation they understood, because if they understood, surely they would not want more procedures and treatments that would mostly only cause more discomfort. But I did not ask. It would be hard enough to look them in the eyes later and tell them dialysis was not on the list of things we could do, I thought as I left the room.

I met my attending in the nephrology fellows' workroom, a small space just big enough for two desktop computers and a dry-erase board for teaching. Dr. Scharf was tall and thin with distinguished wire-rimmed glasses and gray hair. He was our James Brown, the hardest-working man in the nephrology business, usually chewing on a stick of gum so he didn't have to pause to eat. He was a brilliant scientist, having done groundbreaking work in his field, and a gifted clinician, whose understanding of nephrology was unsurpassed. But in response to fellows' errors in clinical acumen or judgment he was known to unleash tirades so cutting they made even the men cry. But just as I was entering the fellowship, the funding for his lab dried up. Now all his time was spent seeing patients. Patients in the clinic. Patients in the dialysis units. Patients in the hospital. Often with a team of medical students, residents, and fellows in tow. And the tirades stopped. For the most part.

I remember when he tried to teach me the *right* way to place a dialysis catheter into the large femoral vein coursing through a patient's upper thigh, a procedure I had done many times and could do with confidence. As we gathered the needed supplies and walked to the intensive care unit where the patient lay mechanically ventilated and sedated, I reviewed the procedure in my mind's eye.

With a Betadine-soaked swab, clean the skin over the vein. Start in the center and sweep in a widening circular pattern. Repeat twice more. Cover the entire patient with sterile drapes, precut so that only a six-centimeter circular area around the target is exposed. I would use the portable ultrasound machine to visualize the vein, just to be extra safe, though I had done it many times by feel alone when no ultrasound was available. Just under the skin above the vein, inject 5 milliliters (1 teaspoon) of lidocaine so the patient will not feel the pain of the larger needles to come. Feel the blood pulse, pulse, pulse through the patient's right femoral artery beneath my left index and middle fingers encased in sterile latex gloves. My right hand holds the needle. The vein is just next to it. Remember which side—there is a V-A-N driving out of the patient's crotch, a mnemonic we learned early in training. Vein-Artery-Nerve. The vein is on the crotch side of the artery. Insert the needle at a 45-degree angle. Burgundy blood oozes into the syringe. This is good. I'm in. Bright red and spurting would mean I had hit the artery—not good. Now move my right hand and rest it against the patient's thigh to hold the needle securely in place as I twist off the syringe with my left hand. The hard part done.

This time would be no different, I anticipated. There would be no ineptitude to provoke a Scharf tirade. We would be done within thirty minutes and on to the next task of the afternoon. Instead, ninety minutes into *his* way of inserting a dialysis catheter, I found myself frustrated trying to balance my hands in midair while holding the needle steady as he instructed.

"This is awkward for me. I can't do it this way. I'm going to do it my way," I announced sternly.

"OK," he said rather sheepishly. "I was just trying to show you the proper way to do it."

"My way works too," I said, already moving along into it.

Silence. No tirade. Perhaps it was my straightforward

manner that protected me from Dr. Scharf's outbursts. I looked forward to being on inpatient service with him, even asked to be assigned to his clinic. As we would wait for the elevator or for the urine to spin down in the centrifuge, he tried to teach me Yiddish words, and in return I tried to teach him some urban slang that, truthfully, I had just recently learned since moving to the Bay Area. While I forgot most of his words right away, for weeks he joked about looking for but never finding a *ho tat* (aka the tramp stamp, a tattoo just above the gluteal cleft, aka butt crack, of a certain type of young woman) when examining his almost exclusively middle-age and elderly patients. *Bless his heart*, I thought.

He listened intently as I presented the case of Mr. Madani. I told him the history of his cancer, the CT scan, the rounds of first-line, second-line, third-line, then experimental chemotherapy, the surgery and procedures, what I found on physical exam, and the lab results. Assessment: fifty-five-year-old man with widely metastatic renal cancer, refractory to all treatment, now in chemotherapy-induced acute kidney failure. I paused before giving my plan—*I think that, therefore, he is inappropriate for dialysis.* He spoke first.

"Better call IR and let them know we need a line now. And be sure to call the charge nurse to let her know she'll need to work him into the schedule for a short run today."

I was stunned. Dr. Scharf's orders conflicted with what I had been taught up until that point in my training about the indications (such as high potassium, ethylene glycol poisoning, and kidney failure with symptoms) and the contraindications for dialysis—like a terminal illness other than kidney failure. Things had changed.

In 1999, the Renal Physicians Association and the American Society of Nephrology published its first edition of the clinical practice guideline *Shared Decision-Making in the Appropriate Initiation of and Withdrawal from Dialysis* to assist nephrologists, patients, and families in making decisions about dialysis. The guide promoted shared decision-making "because it addresses the ethical need to fully inform patients about the risks and benefits of treatments, as well as the need to ensure that patients' values and preferences play a prominent role." Mind you, as a fellow I was completely unaware of this publication. Perhaps I missed the lecture or rounds when it was mentioned, but what a fellow is exposed to during training is more likely a testament to what is valued in an institution, such as an emphasis on *hard* science like identifying the genetic mutation in a congenital kidney disease rather than on *soft* science like improving communication between patients and doctors. But no matter; I don't think my awareness of the publication's existence would have been very helpful in the case of Mr. Madani because there was no conflict that needed guidance to resolve if the nephrologist doesn't question the patient's appropriateness for dialysis and the patient and his family are in agreement. The decision was already made. Initiate dialysis. No committee or even paperwork required.

As instructed, I informed interventional radiology that a catheter for dialysis was urgently needed and wrote the initial dialysis orders. Without question, the radiologist burrowed the plastic catheter tube under the skin just below Mr. Madani's right clavicle and into his chest, where it was threaded into the internal jugular vein. Once the catheter was secured into place with a stitch and taped down under gauze, Mr. Madani was wheeled to the dialysis unit, where the nurse connected him to the machine via his newly placed catheter and carried out my dialysis orders. After a brief dialysis run the next day, he would be discharged to continue dialysis three times each week in an outpatient dialysis unit.

Less than an hour after launching the series of events to dialyze Mr. Madani, I was sitting at a workroom computer when my pager went off. I dialed the number as I sighed and blinked slowly over rolling eyes. *This better not be another consult,* I thought. Unless this was about someone who just arrived in the ER, a request at this hour from one of the primary medical or surgical teams for our team to see yet another patient would be a direct violation of consult etiquette: request consults before noon. Waiting until late in the day when you've known you needed our help since the 6 a.m. lab draw results was grounds for a fellow's snit.

My call was answered after one ring.

"Hello, this is the renal fellow returning a page."

It was the oncology fellow. He had heard of our plan for Mr. Madani.

"Why are you putting him on dialysis? I thought you guys didn't dialyze terminal patients?" There was an undertone of anger and exasperation in his voice, as if I had made his life harder.

"Yeah, that is what I thought too, but my attending disagrees," I replied flatly.

Dialysis would give Mr. Madani more time to get his affairs in order, Dr. Scharf had explained. I remembered the CT scan. Months of chemotherapy. Dozens of oncology clinic visits. I wondered why his affairs weren't already in order. But I had only stood wide-eyed, nodding my head. Fellows did not defy attendings.

*"Why did you guys send him to the hospital?"* I wanted to ask the oncology fellow, but didn't bother, believing I already knew why. They expected us to say there was nothing we could do either. They didn't want to be the ones to send him home to die. They had passed the buck to us, but were surprised and maybe even a little angry when we planned to continue invasive, aggressive care.

Though I could see Dr. Scharf's point and even take in the lesson that there was a role for dialysis in end-of-life care, perhaps as a tool to ease suffering by removing the wastes and fluid that might accumulate and cause a patient to feel nauseous and short of breath. But I did not see how we were easing suffering for Mr. Madani. Rather, we were prolonging his suffering and creating new suffering because with all the good dialysis could bring, plenty of bad would come along too. With dialysis, we would make him leave each treatment more tired than he arrived. We would make him feel light-headed or make his legs cramp from time to time. We would reallocate his time with family to traveling to and from the dialysis unit several times each week. But we would not take away his worst symptom, the constant itching. We were only palliating his fear of dying because dialysis was available, and because apparently saying no is hard, even for the doctors who deal with death so often their patients are called survivors.

Mr. Madani left the hospital as expected, and I went on to the next patient. In the fourth week after my consult on Mr. Madani, his blood pressure became too low for dialysis to remove the fluid that had begun to accumulate in his body without a working kidney to pee it out. He kept going to dialysis because at least it could lower the potassium that, left unchecked, would stop his heart from beating, probably without any symptoms. It was in the fifth week that he showed up so breathless and weak that the dialysis staff sent him directly to the ER. His blood pressure was life-threateningly low. Pressor medications were started to try to bring it up, but they did not. Infection was suspected, so antibiotics capable of killing all known bacteria were started, but they could not kill his bacteria in time.

He had recently decided that he did not want to be intubated in the event he couldn't breathe on his own, so the stiff plastic breathing tube that would have to be slid into his windpipe in order to connect him to the breathing machine was withheld, even though artificial respiration might have given him a few more days to live. Intubation was where he drew the line.

"Some people want to give up at the end. He was not one of them," said the attending nephrologist who inherited Mr. Madani's care in the outpatient dialysis unit. Her choice of words stung me. *Some people. Want. To give up.* As if choosing to stop all the time-consuming and often painful procedures and treatments when moments of time and comfort are dwindling is a sign of human weakness, of failure.

I had assumed that with each failed chemotherapy no oncologist had suggested *No more. Enough.* That they could have moved Mr. Madani off the path of medical intervention, but perhaps it was Mr. Madani who wanted dialysis because he believed it would keep him alive, in spite of his widespread cancer. I wondered what drove him. Hoping for a miracle? Believing that forgoing all that was possible was against God's plan? Not believing any of the doctors? I hadn't taken the time to find out. I wondered if any of us had.

I wondered too what would have happened if I had had a different attending on service that week or if Mr. Madani had been seventy or slightly less resolute. Would the same orders have been given? Would a different set of orders—or no orders at all—have been easier to give?

The original statutory language for the Medicare ESRD program included a requirement for "a medical review board to screen the appropriateness of patients for the proposed treatment procedures," but it was removed from public law in 1978 because the language was thought too vague and neither legislators nor physicians were interested in the government determining patient selection. Besides, they assumed the patient population would remain mostly young and working. They did not anticipate a nation where the average age was over sixty with most people having significant medical problems in addition to kidney failure. They didn't envision a time when questions of appropriateness would arise. And they didn't foresee a time when the Medicare ESRD program would surpass its originally projected cost of $250 million each year. Enacted in 1972, the Medicare ESRD program cost $1 billion in 1979. Most recent estimates are at more than $30 billion.

But here again, individual desire trumped societal good. And Mr. Madani wanted to be alive. By (almost) any means possible.

When I think of Mr. Madani now, I am reminded of a joke I heard sometime after I was involved in his care: An oncologist walks into a funeral home, looking for his patient. The casket is closed. He opens it and is surprised to find it empty. "Where is my patient?" he asks the attendant. "I wanted to give him one more round of chemo." "Oh, they took him to dialysis," the attendant replies.

I laughed at the time, because there is truth in gallows humor. For me Mr. Madani was a lesson in how peculiar medicine can be. It is a place where we fool ourselves into thinking that we can somehow get out of life alive and where dialysis is a means to no end. It is a place where the teaching of dialysis practice becomes more permissive within a two-year fellowship stint, more variable from nephrologist to nephrologist. Dialysis has gone from a miracle to something mundane but to be avoided at all costs to the thing we Americans—no matter if rich or poor, Black, Brown,

Yellow, or White—cling to, because it holds the promise of more time. We—providers across specialties, patients, and families—begin to expect it, feel entitled to it, demand it, and we won't let go of it, even when it doesn't make sense, even when all it really guarantees is more suffering.

# PART V

## LETTING HER GO

# 17

## DIRTY WHITE BOY

Mr. Madani was among the last patients I cared for as a fellow. Clinical training was nearly over. It was time to get a job.

I didn't want just any position. No private practice. No full-time clinical gig. I wanted to be a clinician researcher, where a sizable chunk of my time would be dedicated to doing research. This requirement limited my options, which were further limited by my location.

Avery was only ten years old and, according to California family law, his surroundings remaining the same trumped where his mother lived. So if I left the Bay Area, I would effectively be leaving my baby. Not an option.

I wanted to stay on at UCSF, with the public hospital as my clinical and research base. A faculty position fully paid for by UCSF is a rarity, but one may be allowed to create a position with money from another source, making UCSF an "eat what you kill" kind of academic institution. In order to create my position, two things had to happen to make my wish come true: I had to win the research funding to support my salary and I needed to be invited to join the faculty at San Francisco General Hospital (SFGH). Just two months prior to the end of the fellowship, I had neither.

I knew part of the blame for my unsecured next steps rested firmly upon the fact that, in an institution filled with the best and the brightest, I was just an average fellow in my depth of basic nephrology knowledge. A bit of me agreed with the medical student who was spending a month on renal service and said with a smile, "Oh, no. The kidney is smarter than me," when asked if she was interested in a career in nephrology. While I came to be in awe of the kidney during my fellowship, I went into nephrology because of research interests.

Part of the reason I had no research funding was because I didn't know what I wanted to research that I actually *could* research. Not *could* as in intellectually capable, but as in what the system would allow me to do. I went into the fellowship clear that I wanted to research racial and ethnic disparities—why Black and Brown people were much less likely to get a kidney transplant than White people.

But I was discouraged.

I initially sought out Yin Liu as a mentor back in the days when I was still just looking for a research project and not another tough clinical year. Finding a research mentor is often likened to dating. The junior person, the mentee, may have to go on many dates to find her Dr. Right, the senior researcher who not only believed in her potential but also was willing to commit to helping her realize it.

Our first meeting was in his small, windowless office. Yin was a tall, Chinese American man with a receding hairline and flat, black eyes against a stony expression. I would have felt more comfortable if he were Black or even Hispanic. I would have felt more assured that he would believe in my potential, that he would look out for me. But Black and Brown faculty at UCSF were a rarity, particularly those far enough along in their careers to be mentors. There were no Black nephrologists at UCSF and the one Brown option was about to leave for a position at another institution. But I looked at working with Yin as an opportunity to learn how someone from a different race and culture went about achieving success. What their parents said to them that maybe mine hadn't. What they told themselves that maybe I didn't.

Yin leaned back in his office chair with his left ankle crossed to rest on his right knee. He held in his hands the CV he asked me to bring.

"How did you find me?" he asked with a strong but clear accent. It seemed only his lips moved when he spoke. No jaw. No cheeks.

"I found you on the CRISP website. I was looking for someone at UCSF with an R01 doing research in kidney disease," I said. CRISP, now called the NIH RePORTER, was the National Institutes of Health's searchable database of all federally funded biomedical research projects. I knew that the NIH's R01 grant mechanism, which was awarded to researchers who no longer needed a research mentor, allowed the researcher to apply for smaller research grants to support someone from backgrounds underrepresented in medicine—Blacks, Hispanics, or Native Americans, individuals with a physical or mental disability, or those who grew up in poverty—at every level of education, from a high school student to a college student, a medical student, resident, or fellow.

"That was clever," he said, a slight glimmer in his eyes. I had impressed him. My heart skipped a beat like a schoolgirl whose *Do you like me?* note was passed back from her crush with the *Maybe* box checked.

He went on to interview me. Where I had gone to medical school. How many research papers I had published. It was clear that he was deciding if I was worth his time. Mentoring contributed to promotion and special grant funding, but only if the mentee was productive or, in other words, wrote research papers that were good enough to be published in reputable medical journals.

He agreed to take me on as a mentee.

Like any other mentor-mentee couple, we were happy for a time. We met regularly. He showed me how he approached research as we worked on my project. He took me to the meeting for a huge multi-institution research project he helped lead and introduced me to his peers.

Things changed when I proposed a project for a career development award. A career development award from the NIH provided salary and project support for junior investigators. I proposed to do what's called a mixed-methods research project. I planned to do the usual counting of things as done in *quantitative* research, the most popular type of research. I wanted to count things like how many patients referred for transplant evaluation got kidney transplants—by race. And how many people were involved in deciding who got kidney transplants—by race. But because as Albert Einstein said "Not everything that can be counted counts and not everything that counts can be counted," I also proposed *qualitative* research in order to get at all the reasons why the numbers of who had kidney failure and who got kidney transplants didn't add up. I wanted to interview transplant team members and observe what happens behind the closed doors where the conversations about who will get a kidney transplant are had. Yin perceived my research project idea as too controversial.

"She wants to do a project to show that the transplant system is racist," he said with a roll of his eyes in my career-planning meeting with senior faculty in the division.

"No, I don't!" I protested.

But maybe I did. Because that is how it felt to Robert and me. It felt racist as he sat in his dialysis chair year after year as Whites left for transplant but those who looked like him didn't. It felt racist to me because all the reasons given to explain the racial and ethnic disparities in kidney transplantation, like not being referred for evaluation as early as possible or not enough insurance or social support, were issues that tended to disproportionately affect the poor and people of color. Issues that could be overcome at a system level. Had these problems prevented some of those deemed most worthy from getting kidney transplants, I believe the system would have found a way to overcome them.

Just like it did for dialysis.

But regardless of what I felt or believed, if legitimate research found that a system marginalized a specific group of people in a way that was detrimental to their health and well-being, then it should be revealed. And then the system should be changed.

Nevertheless, soon after that group meeting, Yin said to me alone, "I will not support you if you continue in this line of research."

One cannot launch a successful research career without support. I felt silenced.

"I don't like the way you approach research. It seems biased," Yin said a few months later. Two years had passed since our initial meeting and I sat on the other side of his desk in his new office with a window and twice the square footage of his old one.

I had grown to expect such blunt statements from Yin. His way was to walk through life ripping Band-Aids off without so much as a count to one so a person could brace herself. No beating around bushes. No sugar-coating. I actually liked this about him very much. I found it to be a time-saver, but on this particular occasion his blunt statement left me taken aback.

"I can't help the way that I see the world," I responded. I probably would ask different research questions if I weren't Black or had different experiences, which is the very reason why it is important to have people from different backgrounds doing research. "I think that's where a hypothesis comes from. But then I test that hypothesis with rigorous research methods."

He nodded. "Oh, so you really did interview those people for your project?"

I nodded back before my brain could register why his words burned through my core. He was referring to the study I completed under his guidance in which I assessed the health literacy—the ability to understand written health information—of dialysis patients in five local dialysis units. Though it was a small study of only sixty-two patients (73 percent of whom were Black), I found that dialysis patients with low health literacy had an almost 80 percent lower likelihood of being referred for kidney transplant evaluation than did dialysis patients with adequate health literacy, even after accounting for other factors that might explain why patients are referred or not.

The question was asked jokingly, but just as there is truth in humor, there is also insinuation. And it seemed to me that the implication here was that my research integrity was being questioned. I felt like I had just awakened to a note on the nightstand that read: *Sorry. I can't.—Yin.* The relationship as I knew it was over.

I soon found a new mentor in Ross Pugh, an esteemed internist and researcher who had recently accepted the offer to take on a leadership position at SFGH. Ross was Black, and his research was so focused on kidney disease that nephrologists tended to forget he was a general internist by training. I felt I had won the lottery, but when my prospective boss, Andrew Reeve, was slow to invite me to stay on as faculty, I met with him to ask why.

"Well, the impression is that you take good care of your patients," he said. "Particularly *some* of your patients." His voice trailed off and he looked away.

I thought back to when I interviewed for the fellowship program. I was forthcoming about my essay about my experience of becoming a kidney donor. It had been accepted for publication, I told them. They smiled and nodded and soon I received a letter of acceptance.

"Good for Harvest, Bad for Planting" was published shortly thereafter. It led with my experience as a resident at Highland, where it was clear that our poor, mostly of-color patients were a popular source for harvesting organs but disproportionately lay fallow when planting time came.

The piece resonated with people widely. Its spread felt viral. It brought some people to tears. Others were enraged at the problem. There was a general outpouring of very positive comments from the general population and even primary care doctors.

But there was a stillness as the circle approached providers with ties to the kidney transplant system. A blog among a group of transplant social workers was incredulous. They wondered in their posts, how could it be so? Among nephrologists and other transplant team members there was complete silence, even though I circulated the piece myself to several people in my new program.

"Most people are taught that if they don't have anything

good to say to not say anything at all," Yin had shared with me back when he told me that he would not support my research project idea to examine why kidney transplant disparities existed. He said it in usual form, bluntly with a roll of his eyes. "People felt it was unfair. That you were shooting from the hip." And the fact that I had e-mailed it directly to several people in my new program was viewed as insulting. Like I was rubbing it into their faces, he explained.

I honestly thought it would be an impetus for reflection on the transplant system. It hadn't occurred to me that others would take a piece about *my* personal experience personally. How naive of me to think people could be bigger than their egos, but surely they couldn't think the system was perfect, could they? Surely they could see that the closed doors, the secret lists, the data on who got transplanted and who didn't bred a certain amount of distrust. Even from a doctor. Couldn't they?

While health-care providers may acknowledge that racial disparities in health care exist, most tend to believe they aren't part of the problem. Apparently many within the local kidney transplant community felt the same. They were offended that I had the audacity to blatantly accuse them of being no different from all the other kidney transplant programs around the country that they believed were the ones really responsible for disparities. There seemed to be no room for reflection. Instead it appeared that not only had they taken my experience of feeling discriminated against personally, but they also tried to make the case that it was me who was the discriminator.

"She made a rookie mistake," a Black transplant nephrologist at another institution who was sympathetic of my predicament explained to a colleague, because even though I gave no names, everyone knew the identity of the transplant nephrologist I wrote about. The kidney transplant world is a small pond and I had called out a big fish.

I was proud of my essay, even though, admittedly, some of my analysis wasn't 100 percent on point. For example, when I wrote the essay, I was outraged that Robert had been on dialysis for so many years without a transplant. I've since learned that at some transplant centers the wait for a kidney could be eight years, especially for people with blood type O. I also learned that where you lived had a lot to do with when you could expect a kidney—at a transplant center just ninety miles away from us, the longest wait time was five years. From my perch as a primary care doctor, I thought that level of kidney function factored into who was offered a kidney transplant, with the neediest given priority, as it is with liver transplants. I've since learned it has everything to do with if a person with advanced kidney disease knows they have advanced kidney disease and has access to a nephrologist who thinks they are a good candidate for transplant, and the nephrologist actually refers them—a cascade of requirements vulnerable to the effects of personal bias and racism at an institutional level. Anyone who says differently is either blind or lying to himself.

A few years ago, during a discussion of my essay in a medical student writing workshop I led, one student asked with a *gotcha!*-yet-curious lift of his left brow and squint of his eyes, "Now that you're on the inside, do you still feel the same?"

"I *get* it . . . but it's still bullshit," I responded without pause. But once the smiles and laughter that erupted around the room died down, I repeated my more thoughtful, complete rationale. "Because if the reasons they give to explain disparities in access to kidney transplant kept the 'worthy' from getting one, they would have figured out ways to overcome them. Just like they did for dialysis."

"Besides," I went on, "even as a nephrologist, I still didn't know what happens behind those closed transplant allocation doors."

I gave my kidney to help Robert overcome the trappings of the transplant system. The essay was my way of helping those who were still trapped. I couldn't sit silently. Even if it did cost me a job.

I happened to be assigned to the hospital service with Andrew at SFGH when I asked to meet with him about staying on as faculty. It was my opportunity to show him firsthand that I was a good choice to join his nephrology division as faculty, in that I was smart enough and willing to take excellent care of *all* of my patients.

It was my last day before leaving SFGH for my final rotation as a fellow at UCSF. It was my habit to say good-bye to patients. I felt it was bad enough that the doctors were always changing; the least we could do was soften the transition by making a point of saying good-bye.

Room after room, Andrew and I visited the patients whose care we were involved in.

"So I wanted to let you know that this is my last day on service here, but Dr. Gupta will be taking my place. She is very good and I will tell her all about you," I said to each one, and they would smile and nod their own good-byes. Except the last one.

He lay bare-chested in his hospital bed, making the DIRTY WHITE BOY tattoo covering his entire chest impossible to go unnoticed.

"Oh, no!" he exclaimed, as he threw his hands up then slapped them down on the bed. "You're the only one who tells me the truth!"

Out of the corner of my eye, I saw Andrew's eyebrows lift and his forehead wrinkle.

# 18

## THREE LADIES

Not long after I joined the faculty at San Francisco General Hospital I met eighty-three-year-old Ming Lee in the pre-dialysis clinic where I served as the attending nephrologist when my colleague was unavailable. Just a few months prior, my colleague was one of my attendings and had a good three decades of nephrology experience over me, so I tended to not change the care plans already under way for patients returning to the clinic.

Mrs. Lee's plump face was constantly smiling. *OK. Thank you. Doctor.* These were the only English words, usually strung together, I ever heard her say. A Cantonese interpreter translated the rest and her granddaughter sat silently beside her. Mrs. Lee had an eGFR of 16 milliliters per minute, only about 3 teaspoons of blood coursing through her kidneys' filters each minute, when normal for her age is closer to 17 teaspoons. She also had an extensive list of other medical problems—blockage in her heart arteries, ministrokes, gout, and a recent bowel obstruction—just to name a few. She was scheduled to see a vascular surgeon the following week to talk about creating a fistula for hemodialysis, but she was ambivalent about going to that appointment and about dialysis in general.

I was ambivalent too. The plan for Mrs. Lee felt wrong. Given her age and overall health status, I didn't think dialysis was the right thing to do.

Though I wasn't aware of it at the time, a small but growing body of research supported my intuition. It showed that patients similar to Mrs. Lee—over seventy-five and with serious medical problems in addition to advanced kidney disease—were as likely to live as long *without* dialysis as with it and often with a better quality of life. This research comes mostly from the United Kingdom, where about 15 percent of elderly patients with end-stage kidney disease die without ever starting dialysis. They have programs in place to provide *conservative management*—treatment aimed at minimizing symptoms of kidney failure while maximizing the quality of life remaining without dialysis.

The United States doesn't track patients who don't start dialysis, but almost all elderly patients who start dialysis do in-center hemodialysis, and patients over age seventy-five are the fastest growing group starting dialysis, their numbers having doubled over the last two decades. The burdens of hemodialysis—symptoms of extreme tiredness, cramping, and dizziness; dialysis access–related procedures; and travel to and from the dialysis center—are common among all patients, but particularly so among elderly patients. A 2009 study published in the *New England Journal of Medicine* showed that almost two-thirds of elderly nursing home patients were in worse shape—either less able to take care of their own basic needs or dead—within just three months of starting dialysis, suggesting that treating such patients with dialysis was in direct violation of one of medicine's guiding principles: *primum non nocere*, first do no harm.

Like most nephrologists just completing training, I felt well prepared to diagnose acid-base disorders and do a kidney biopsy, but completely unprepared in how to talk about or practice conservative management. Rather, I was taught that transplant is better than dialysis and that dialysis is better than death. Always.

It was as if dying of kidney failure wasn't allowed. As if the fact that dialysis existed and was readily available automatically meant people should never die from kidney failure. We approached dying of kidney failure as a never event, a tragedy akin to dying of colon cancer because a colonoscopy wasn't done, or of cervical cancer because of missed Pap smears. *She died of kidney failure? How could that be? We have dialysis!*

But would it really be such a tragedy if Mrs. Lee did die from kidney failure?

Most people say they want to die in their sleep, presumably from a heart that simply stops beating. Maybe dying of kidney failure after you've lived a long life is a close second to dying painlessly in one's sleep. I hadn't witnessed it at the time but I've since learned that the person becomes progressively sleepier and sleepier. Over weeks. Months. Even years. So slowly, they may not be fully aware that they are sleeping more. Some have nausea and shortness of breath. Some have muscle cramping pain. But nothing that couldn't be tempered with a little shot of this or that—until they just didn't wake up anymore.

As a fellow in the pre-dialysis clinic, what I observed for how to have discussions with patients approaching end-stage kidney disease seemed consistent with the monolithic, unquestioning agenda implied by the clinic name. The response to patient hesitation toward dialysis or outright refusal of it often felt threatening, coercive, even bullying to me.

"Start dialysis or you'll be dead in two weeks."

"You have a responsibility to your grandchildren to be here."

"If you refuse to start dialysis, then you will be discharged from this clinic."

These were the refrains left in my mind. I was determined not to repeat them. I was no longer the fellow just doing what I was told to do, saying what I was told to say. My actions were my own. My words were my own.

I tried something different with Mrs. Lee that day.

"Not everybody chooses to start dialysis," I said delicately, tiptoeing into a conversation about the possibility of another course.

But before the interpreter could say my words in Cantonese, I watched the granddaughter shift in her seat and the cross of her arms tighten. Though she hadn't spoken a word, it was clear that she understood English and that she didn't like what I was saying.

I didn't know how to move Mrs. Lee off the path she was on without making her or her family feel that I was denying her care or sentencing her to death. I didn't know what a different path would bring.

I retreated.

"Well, maybe you should just go to the appointment and hear what the surgeon has to say and then decide if you want to go forward with it," I said.

With these words, the granddaughter's posture softened and Mrs. Lee smiled and nodded. *OK. Thank you. Doctor.*

Five months and two surgeries later, I saw Mrs. Lee in clinic again. She had a fistula buzzing in her left upper arm and I said nothing to suggest she consider a path that did not involve dialysis.

Another year had passed when I learned that her kidneys had failed to the point that my colleague thought starting dialysis was appropriate. Her fistula was ready to use. She started dialysis with a Cantonese-speaking nephrologist in a Chinatown dialysis unit.

*Oh good,* I thought when I heard the news. She would be with people who spoke her language. A community who understood her experience.

A few months later, I walked into our renal center administrative office. I checked my mailbox. I signed off on orders for a new patient's peritoneal dialysis supplies. I chatted with office staff. Like any other day.

"Do you remember the patient Ming Lee?" asked Bao, the office administrator. Chinese American Bao was a dialysis technician before joining the business office. She sometimes worked as a dialysis technician at local dialysis units, both to help out and to keep her skills sharp.

"Yes, of course. How is she doing?" I asked brightly.

"She jumped off the roof of her five-story apartment building."

*She. Jumped. Off. The. Roof.* The words hit me in the chest so hard they took my breath away.

I imagined how unheard she must have felt, even in Cantonese. How dark her world must have been, with no sign that the sun would shine. Ever. Again.

Maybe she jumped off that roof because of something completely unrelated to dialysis. But maybe it *was* all about dialysis and she saw no other way out. And I didn't have the training, the words, the courage to show her there was. I tried not to cry.

I never wanted to feel like that again, so I wrote about the experience with Mrs. Lee in my application to win one of the sixty-four spots in Harvard's yearlong Palliative Care Education and Practice (PCEP) program. The goal of the program was to teach clinicians how to better deliver palliative care to patients with serious illness and how to teach it to other clinicians. According to the World Health Organization (WHO), palliative care is "an approach that improves the quality of life of patients and their families facing problems associated with life-threatening illness, through the prevention and relief of suffering by means of early identification and impeccable assessment and treatment of pain and other problems, physical, psychosocial and spiritual," so a good chunk of the on-site PCEP time was devoted to lectures with titles such as "Hope" and "Pain" and "How to Take a Spiritual Assessment." Definitely not the usual stuff of our renal grand rounds, where lectures with titles like "Angiotensin II Type-2 Receptor Signaling and Renal Sodium Handling" and "Molecular Approach to Sensitized Kidney Transplantation Recipients" were the norm.

My peers were mostly physicians, but there were quite a few nurses and a smattering of pharmacists and social workers too. Since its inception in 2000, there usually was one participating nephrologist each year. I was the nephrologist in the class of 2014. While I welcomed the strange new concepts that PCEP presented, I spent my time trying to translate everything from the program's oncology focus into something that made sense in nephrology. It felt impossible, not because cancer is more serious than kidney failure—kidney failure has no cure either and dialysis patients die at a rate nearly twice that of many cancer patients. Rather, I found it impossible because of the knowns of oncology that simply weren't the case in nephrology. In oncology, it is known with some certainty how long a person can expect to live depending upon the kind of cancer and how far it has spread from where it started. In oncology, it is more or less known what side effects a particular chemotherapeutic agent will have on patients in general.

In nephrology, dialysis creates a limbo. It is difficult to predict exactly how dialysis will affect the individual patient. Some who we expect to be miserable on dialysis do just fine.

Some who we expect to do just fine suffer. Though some dialysis patients die faster than cancer patients, some live decades on dialysis. Some would live just as long without dialysis as with it. And some would die within days without it. So because we are not sure, we tend to start dialysis by default.

When I met Mrs. Durante in the pre-dialysis clinic, she was spending most of her time in a wheelchair since a long, complicated hospitalization three years prior left her there. The long hospital stay was followed by an even longer stay in a nursing home, which she said felt like torture and nearly killed her. She never wanted to go back.

She was only sixty-three in chronological years, but as I read her medical record and looked at the frail, listless-appearing woman before me, biologically she seemed closer to her mid-seventies. She'd had strokes. She'd had surgery on the bones of her neck to release the hold they had on her spinal cord. She had heart failure. Diabetes. So when she sat facing the fork in her road to complete kidney failure, she was unwavering in her refusal of dialysis. She'd had enough, she said.

Perhaps dialysis would not be beneficial in terms of adding quality or length to her life, I said, and proposed conservative management of her kidney failure. We could treat the symptoms of kidney failure as they arose without dialysis, accepting that eventually she would pass from kidney failure—if one of her many other health problems didn't take her first.

Her pale blue eyes widened in surprise as if she expected me to argue with her, perhaps as the nephrologists who had seen her before me did.

*Don't say that. You will die without dialysis.*

*I'd rather die than go through that.*

*Don't say that. It won't be so bad.*

*No, no, NO! I don't want it.*

Instead, I had inadvertently called her bluff. Her poker face was exposed. *Well, maybe a little dialysis,* she conceded.

She chose peritoneal dialysis and her daughter Josefina would help her do it. Jo was the youngest of Mrs. Durante's three daughters and completely devoted to her. I agreed that peritoneal dialysis was a perfectly rational choice too and was happy to support her in that decision as well.

I had softened from my original all-or-nothing way of thinking about dialysis: *If very old or very sick, then don't start it. If already on it and the patient becomes very old or very sick, then stop it.* But the problem with all or nothing is that most in my experience tended to opt for all in the off chance or even delusional hope that all would restore the body to where it was before everything fell apart, or at least prolong its existence for a few more weeks, days, hours.

I've learned to think about peritoneal dialysis as an alternative to having to choose between frightening unknowns. Yes, peritoneal dialysis would involve a catheter being threaded into the space between skin and the peritoneal membrane covering the abdominal organs. Yes, the skin around the catheter or the peritoneal membrane might become infected, causing pain and sometimes fever and chills, but this is an infrequent occurrence. And, yes, the frail elderly patient would need a family member to do the required handling of dialysis fluid bags and connecting them to the catheter, but the dialysis itself would be much gentler.

There would be none of the dramatic changes in body weight of in-center hemodialysis as a result of liters of fluid being siphoned off in just three to four hours three times a week (because in a body unable to pee it out, the fluid only reaccumulates between siphonings). There would be none of the inevitable exhaustion after dialysis or cramping during dialysis. With peritoneal dialysis, the siphoning is slower and of smaller amounts over the course of every day. And peritoneal dialysis can be tapered off gradually without the fear of rapid death triggered by an in-center hemodialysis treatment or two.

In a 2008 study published in the *Clinical Journal of the American Society of Nephrology*, researchers found that if dialysis providers answered "no" to the question "Would you be surprised if this patient died in the next year?" the patient was 3.5 times more likely to die within the year. Given Mrs. Durante's desires to avoid interventions and my "no" response to the "surprise" question, I employed what a few colleagues and I have termed a palliative approach to dialysis care. My goal was her goal—to get her feeling better. Lessen the nausea. Improve the fatigue. I didn't worry about achieving the standard metrics that defined quality care for dialysis patients but made her feel worse. For example, a blood phosphorus level of less than 5.5 was the standard of care, but the large pills that bound phosphorus in the stomach before it could be absorbed into her bloodstream—phosphorus binders—made her gag. I told her that her phosphorus of

6 or so was OK. Sure, rigorous adherence to dietary phosphorus restrictions would have minimized the need for the binders, but I told her it was OK to have a bit of her favorite ice cream. After all, the intention of the goal and other standard metrics was to minimize complications that happened over the long term and maximize survival. Mrs. Durante, I knew, did not have a long term. Her goals and mine were to maximize how well she felt and how much she could enjoy life. Sometimes that meant minimizing the parts of dialysis care that got in the way of those goals.

Had she chosen hemodialysis, I would have taken the same approach. I would have blatantly disregarded the mantras of "fistula first" and "catheter last." Sure, pushing her to have a fistula created or a graft put in certainly would have decreased her chance of getting an infection, but it also meant getting stuck with two needles three times a week and painful procedures required more often in older patients to keep it working. If she always started cramping in that third hour of dialysis, I would have stopped it at two, even if that meant her blood pressure was a little higher than it was supposed to be because we didn't have time to siphon away all the extra fluid in her body. Even if that meant the blood test said she wasn't getting "adequate" dialysis. I would have stopped getting so many blood tests in the first place.

Many if not most of my colleagues would not agree with me. They would accuse me of substandard care because they believe in the goodness of dialysis and that any downsides of it are worth it. They believe in one size fits all. Palliative care is equated with giving up, something you resign your patient to when there is nothing else left to do.

And then there's the money.

We get paid much more to keep someone on dialysis than to keep them off of it. If we don't achieve dialysis metrics—like avoiding dialysis catheters or providing a certain dose of dialysis—known to best result in long-term benefits, we are financially penalized. But create a fistula in a little old lady that usually requires interventions to make it work and keep it working and make her stay on the dialysis machine as long as it takes for the numbers to look right, then essentially get a bonus. If we see an in-center hemodialysis patient four times in a month, we stand to make 50 percent more money than if we only saw her once. And the nephrologist really only has to see the patient once each month—if a physician assistant sees the patient the other times, we still get paid. We would have to document a *comprehensive* medical history and examination over the better part of an hour with a patient returning to clinic *twice* to see the same money—and good luck trying to justify why that was clinically necessary to do. The second, third, and fourth in-center hemodialysis patient visits can be more like drive-bys—a simple documentation that we (or the physician assistant) "saw" the patient, with no notation of time required. Private insurance companies and the Medicare ESRD program pay top dollar for dialysis care, not clinic visits. It's profitable to build another dialysis center, but we haven't figured out how to build comprehensive outpatient palliative care services.

Not that every nephrologist starting a frail, eighty-something-year-old with failing kidneys on dialysis is doing it for the money. Many would truly believe that is the right thing to do. But as one nephrologist said, "I'd dialyze a tree if they paid me for it," I can't help but think that if thoughtful, prolonged, and repeated conversations with patients and families about the realities of end-stage kidney disease and dialysis were financially incentivized instead, we would be less quick to put and keep people on dialysis when it is unlikely to benefit them in terms of quality or length of life.

Despite my efforts, Mrs. Durante continued to have some discomfort. In her first few yards down the path of peritoneal dialysis, she encountered peritonitis. The skin around her catheter and her peritoneal membrane became infected. She had pain and fever and chills. Her shoulders slumped and she looked up at me with a sigh that said *You're making my life worse with this.*

"Hang in there," I encouraged, and with appropriate antibiotic treatment, she was able to get past it.

Her nausea never fully went away. She had trouble swallowing even the chewable and powdered forms of binders. And it took a while to calm the painful itchy rash in the fold beneath her hanging belly.

All that said, what was happening was not so bad compared to what she had already come through, and she was able to feel well enough to begin enjoying life again. She and Jo were even able to go to the opera regularly. There appeared to be weeks of happiness between bouts of her various ailments.

But two summers later she needed to be in the hospital—for weeks. She became weaker and weaker. She used to be able to push herself to a stand using the arms of her wheelchair, take two steps, and sit down again in a different chair. Now such a feat would require at least two strong-backed bodies to make it happen. She had to face the reality that going home, at least for a while, was not an option. Skilled nursing facilities could not do peritoneal dialysis, so not going home meant she would have to switch to hemodialysis.

Going to a skilled nursing facility again and doing hemodialysis. Two things Mrs. Durante had said she never ever, never ever ever wanted to do, so I began talking with her about the option of stopping dialysis. What symptoms she might have if she stopped. How much time she might have. On average people with no kidney function who stop dialysis die in eight days. She had a little kidney function left, so she probably would live a bit longer. I didn't know how much longer she might live if she continued dialysis. I only knew it would be hard.

It is our inability to accurately predict when people will die that usually keeps us from preparing patients for death. In one survey of dialysis patients published in the *Clinical Journal of the American Society of Nephrology* in 2010, less than 10 percent reported that any doctor had ever discussed prognosis with them. It's not surprising that we nephrologists are uncomfortable with diagnostic uncertainty given that nephrology is a field rife with equations. There are equations to calculate how much water the dehydrated body needs, more equations to calculate what percentage of the body's sodium is being peed out, and still more to calculate how well the kidneys are filtering. Somehow we are able to talk about these calculations freely and confidently even though they are just estimations. While tools to estimate prognosis among dialysis patients and tools to estimate prognosis among patients with advanced kidney disease are in development, without a crystal ball it is doubtful that any tool will ever have enough precision for nephrologists to feel assured of accuracy for the patient before us.

However, since the vast majority of patients and families only have *their* experience with illness up to the present moment, our clinical knowledge and experiences with similar patients about what the future may hold are invaluable—and should be shared. To ask patients and families to make decisions in the abstract, with no knowledge of what might come of them, is at best cowardly of us nephrologists. At worst, it is a clear demonstration of detachment, a sign that we are not in it with them. We say we want people to make decisions for themselves, but mostly I think we are afraid of being wrong, afraid of being held responsible, afraid of being sued.

Besides, most of us got into this business of doctoring because we wanted to help people. What has always been the implied indirect object here is "live." We got into this business to help people *live*. To switch that out to "die" is counter to how we've been raised.

Yet published research studies suggest that patients and family members want to be given information about life expectancy, even if prognosis is poor. Others have shown that those engaged in shared (as in with input from their doctor), informed decision-making are more likely to make decisions about dialysis and end-of-life care consistent with their personal values—often resulting in preferences for less aggressive care and more conservative management.

This didn't feel like my experience with Mrs. Durante. I tried to figure out her goals and align her care with them, but it is interesting what we become willing to put up with when the universe calls our bluff on *I'd rather die*. Then all that really, really mattered to her was being able to enjoy food.

So her peritoneal dialysis catheter was removed and she was transferred to the nursing home with a hemodialysis catheter in her chest instead. Then, in and out, in and out from the nursing home to the hospital she went every few days for the next couple of months. She was short of breath. She had chest pain. She was dizzy. Her bottom burned with deep ulcers and uncontrollable diarrhea. Her time was spent staring at the walls or the television. She lost her appetite and wouldn't eat. She refused to work with physical therapy. The possibility of ever getting strong enough to go home was becoming less and less a realistic hope.

I went to see her in the hospital and broached the subject of stopping dialysis again.

"Remember when you said you always wanted to be able to enjoy food? You aren't eating anymore," I reminded her. Soon even when she tried to eat, she couldn't reliably swallow without choking. Her food needed to be pureed.

"I am suffering," she acknowledged. "All I do is stare at the walls. There is no enjoyment."

"We can always change what we are doing," I said.

"No," she said after a few days. "I will keep going." Enjoying food, she had decided, wasn't that important anymore. She had even convinced herself that the pureed turkey she had for Thanksgiving was good. Then what really, really,

really mattered was being around to see her first grandchild born.

"Is anybody pregnant?" I asked Mrs. Durante, trying to assess how realistic this new goal was. I was prepared to apply my newly acquired PCEP skill of asking *"And what else"* until we arrived at a goal that we realistically had a shot of achieving.

One of her daughters was married and trying to get pregnant.

"OK." I smiled. "That sounds like a wonderful goal."

Mrs. Durante was discharged to a nursing home but did eventually make it back home and even back to peritoneal dialysis. Then in and out, in and out, this time from home to the hospital she went with shortness of breath that doctors thought might get better with more fluid removed with a little extra dialysis, a breathing treatment, or maybe some antibiotics. Sometimes she went to the hospital with dizziness that doctors thought might lessen with a little more fluid given back. She was never home for more than a couple of days. But I had stopped trying to have conversations with her about how she hoped to live out the rest of her life.

For every five of our dialysis patients, one will die within the year. Yet we often don't bother spending the time it takes to have conversations about what patients would want their care to be like at the end of their lives. Since dialysis can be a life-saving treatment in many circumstances, we develop a false sense that sudden bouts of illness serious enough to land our dialysis patients in the hospital are temporary when, truth is, dialysis cannot change the reality that the path of kidney failure is a continuous one toward death. A path that is littered with sudden illnesses and setbacks, and recovery is never back to the level of function that the person enjoyed before.

It wasn't that long before Mrs. Durante was back in the hospital, sicker than ever. But all the other doctors behaved as I did and as we tend to do, as if the acute illness she was facing in the moment was temporary. As if her goals of seeing the grandchild five months away from being born or just going home again were still realistic.

I think some of problem is that we—clinicians, patients, and families alike—get caught up in a person's chronological age. We believe that eighty is the new sixty and that only death after at least a century on the planet is acceptable—even if the body has been aged before its time by hard living involving drugs or alcohol or by just plain bad luck of having blood vessels prone to clotting or cells prone to transforming into cancer.

In the end, Mrs. Durante, the woman who wanted nothing to do with hospitals and nursing homes, died in a hospital bed after months upon months of hospital admissions and nursing home stays. She died in a flurry of tubes, compressions, electricity, IV medications—and dialysis. She died surrounded by the code team of doctors and nurses while her children paced and worried and prayed in the waiting room, not realizing that what was happening was not temporary in the way they hoped.

"I don't know what the hell happened," Jo said when I called her a few days later to express my condolences. There was that familiar tinge of anger that families take on when they need to find fault. Someone to blame, because she never would have died, they believed, had it not been for those stupid, insensitive doctors. So what that she had been in the hospital more than not in the last year. She wasn't supposed to die. Not yet.

I had wondered how much Jo factored into her mother's willingness to endure when all that really mattered was no longer possible. Jo's entire life had revolved around her mother. Long before we got to this point, I tried to meet with Jo alone to try to help her come to terms with the reality of what her mother was going through, but something seemed to keep coming up to prevent her from coming by when she said she could.

"I think she was dying," I said softly. "And sometimes, no matter how hard we try, we can't change that fact."

It could have gone the same with Mrs. Nisnisan. She was ninety years old when she died. The last time I saw her in pre-dialysis clinic, she was slumped in her wheelchair as if she had no bones. When I met her two years prior, her kidneys filtered about a teaspoon of her blood per minute. Now they filtered less than half of that and all the blood test results related to her kidney function were so far from normal, so ugly. Ugly numbers make doctors and nurses anxious. They make us want to do something to make them go away.

We could have admitted her to the hospital. We would have been completely justified—the patient was dying of kidney failure. We could have placed a dialysis catheter in her chest and connected her to a dialysis machine until the numbers were pretty again.

And if removing the wastes and extra fluid didn't improve her breathing, we could have pushed a breathing tube down her throat and connected her to a breathing machine to do that for her too. We could have inserted an IV line or two to start the medicines to try to bring her blood pressure up to a more attractive level. At least 90.

We could have even done chest compressions, not so hard as to break too many ribs, but hard enough to keep her heart pumping and just until we could get the noninvasive pacemaker pads in place. That way we could have sent enough electricity through her chest to try to make her heart beat a lovely sixty times a minute or so.

That way we would be able to say words like *everything* and *fighter,* as in "We did everything" and "She was a fighter."

Instead, we did everything to help Mrs. Nisnisan not suffer. We stopped doing everything that could hurt her. For months we had been taking away medicines that weren't helping her feel better. There would be no more sticking her for blood tests.

She slept more and more of the days away until one day she stopped talking and eating in her awake moments. Then her breathing slowed and her breaths shallowed until she stopped breathing altogether. At home. Lying in her bed. Surrounded by all her children. Just how she wanted.

# 19

## WHAT LIES AHEAD

It's easy to talk about courage and letting go and untethering from dialysis in the abstract. It gets much harder, bordering on impossible as we get closer, when it's about you or the person you love. Even though I gave Robert a kidney to minimize his burden and maximize his survival, it's still end-stage kidney disease. Transplants tend not to last forever. Reflecting back on an event that happened not quite six months after our wedding, I paused to ask myself *What would I do to prevent death from happening if it were Robert?*

We were on the phone one evening, and I had just gotten Avery off to bed. Robert was in Boston, having just returned from visiting for the holidays. While our wedding was the first of Robert's plans after transplant and years of being afraid to make plans, Boston was part of a close second on the list—getting back on the career path dialysis had forced him off. He had enrolled in graduate school at Harvard's School of Public Health with the intention of getting a doctorate and left within weeks of our wedding.

"I need to see the optometrist soon. My vision is really blurry," he said. He paused. I could hear him gulping in the background. Ever since we got past the transplant surgery hospitalization, I loved watching him guzzle down whatever ice-cold drink he was craving. He paid so dearly for giving in to his cravings when he was on dialysis. Then there would be swollen legs and face and maybe even shortness of breath until he could get tethered to the dialysis machine to siphon the extra fluid away, but because so much was being removed so fast, cramping was sure to come. But after transplant, there would be no more swollen parts or shortness of breath in response to his guzzling. The urge to pee was soon coming to make everything all right again.

I almost smiled at the thought until he came up for air and gasped, "Whew, I'm *so* thirsty!" and the pit of my stomach twisted with a new thought: *Diabetes.*

New blurry vision. Excessive thirst. These were two of the major symptoms of newly starting diabetes. I knew Robert was already at risk because he was a bit overweight and his father had diabetes, but he had just had his transplant checkup during his visit home and all of his blood test results were fine. What neither of us knew was that one of his transplant medications placed him at further risk.

"I think you need to go to the emergency room," I said, worried.

"What are they going to do for me there that I can't do

for myself here?" Robert asked. He had an aversion to asking for or even accepting help, but his attitude toward doctors had taken a turn for the worse since the days of Matty Kravitz, the nephrologist who could never remember having met him.

"Uh . . . they could give you insulin," I said. "It sounds like your blood sugar is high. There's nothing you can do at home to make it come down."

And then he said something crazy on the order of "I was thinking about moving that sled into the living room."

Now I was afraid. "What did you say?"

"When?"

I felt a pang shoot through my core. I tried not to panic.

"Robert, please. I need you to call 911. You're not making sense now. I'm worried."

"OK, OK. I'll go," he gave in.

We hung up and my mind went into overdrive. *Should I go to Boston? Of course I should go to Boston. But what about Avery? I'll take him with me. I have to buy plane tickets. What will I find when I get there? Oh, Lord Jesus, please let him be OK.*

I feared my wonderful, intelligent, but often stubborn to his own detriment husband had waited too long to go to the hospital. I envisioned standing with Avery a few feet behind the police with a battering ram breaking the door to his apartment down to find him comatose on the couch.

My prayers were answered. Robert made it to the hospital conscious after driving himself to the emergency room. He drove through eyes that could barely distinguish how many fingers someone was holding up and with a quart-size plastic bottle of red Kool-Aid on the passenger seat to drink along the way. He was *so* thirsty! How does the saying go? God looks out for babies and fools. It definitely applied here.

By the time Avery and I arrived at the hospital, Robert was in a private room. He was pissed. Not only had he spent a full day in the ER on a gurney, but his nurse was late bringing him his antirejection medications even though they took—and lost—his pill box in the ER. And she wouldn't give him any information. And a resident physician stuck him four times trying to get blood from the artery in his wrist. And his local nephrologist had yet to come to see him. It made him flash back to how he felt when Kobayashi never came to see him after performing the surgery that caused him so much pain. He felt abandoned. Unimportant. Again.

But mostly Robert was angry because he had this new diagnosis of diabetes. It was another major thing to deal with, another example of his body failing him. And so soon after the transplant that was supposed to fix it all.

Maybe it was just the stress and shock of it all that made him say the words that brought me to tears.

"Man, if I lose this kidney, that's it. I'm not doing dialysis again."

"*What?!*" I said.

"I figure if the kidney fails, then the universe is trying to tell me something. I'm just not supposed to be here."

My words *Let's make a baby,* and his response, *I can't do that,* flashed through my brain. In my mind, transplant was still the panacea to make it all right then. To make it OK to go down the path of marriage and 2.5 babies and a house with a white picket fence and a minivan parked in the driveway.

"Don't say that," I said, not knowing those were the exact words his mother had uttered two decades earlier as they walked into the pediatric nephrologist's office to first learn his fate. Tears streamed down my face and Robert just looked at me with the usual *Uh, water is coming out your eyes* look on his face.

I don't remember what happened next. Robert says I stormed out in a huff, Avery in tow, and didn't return until the next day. By the next day he was feeling better, he had his medications, and his nephrologist had come to see him. And I had chosen to push his words down into the recesses of my mind, pretending they had never been spoken.

It was not until we were approaching our tenth transplant anniversary that I brought them up again. By then I was a nephrologist. One who preached about patient wishes and having the courage to let go and whatnot. One who had seen and heard and felt what happens when we refuse to let go.

As we recounted the scene, tears began to stream down my cheeks just as they had before. I cried because I knew that even though it would hurt me, one day I might have to let him go. Untether.

If that day comes, I will honor what Robert has said if he is unable to speak his wishes for himself. I will not push him to do what his spirit can no longer endure—because his words that day in Boston were from a position of real knowledge, from someone who *had* been through dialysis. For nearly six years. There was nothing abstract in his words.

He is clear, so I will put his needs before mine. I will be selfless in a way that I couldn't have been ten years ago in that Boston hospital room had things taken a turn for the worse. Selfless in a way that I wasn't even with giving Robert my kidney because I clearly got something out of it. I got the likelihood of a longer and better future with him than dialysis could promise. I got to feel good about myself. I got to forgive myself a little for past mistakes.

But I hope my resolve is never tested.

I hope that the kidney I've given Robert will last another forty years and counting, just like the kidney given to The Other Robert Phillips. That in the end we die together in our sleep, hearts slowing to a stop at the same time.

But if it doesn't last until then, I hope that maybe a second transplant won't be as hard to come by as I fear it will be. Like anyone who has had blood transfusions or a pregnancy, no doubt Robert has developed new antibodies that will attack anything foreign. But researchers are perfecting the process of desensitization in which they filter antibodies that might attack a new kidney out of a patient's blood, making it easier to find a match for people with a lot of antibodies. Or maybe the surgically implantable artificial kidney that researchers at UCSF and Vanderbilt University have already been working on for more than a decade will be ready in time.

Sometimes as I lie against him on the couch watching television, his left arm around me, I feel his fistula buzz, buzz, buzz and think maybe he would change his mind and restart dialysis if we found ourselves in a place where all our other hopes fell through. After all, he could have had a surgery to take the fistula down.

But he hasn't.

He says he hasn't because he doesn't want to jinx the kidney. I say it is our backup plan. Our lifeline.

Recently I came across a box of keepsakes with a love letter that Robert wrote to me in it (with a little help from Winston Churchill's letter to his darling Clementine) from May 19, 2005, a little more than a month after our surgeries. It read:

*Vanessa,*

*I know that neither one of us had in mind a year ago that we would meet someone and fall in love, but it has happened. And for that, I have no regrets. In fact, it is one of the best things that happened to me in years. For this, and what has happened to us and between us, I have you to thank.*

*For the past year, you have brought so much joy to me that words can never explain. In the past year, you have brought so much life back into a lifeless body; I know and realize what it is to love and to feel loved.*

*You have so enriched my life. I always feel so overwhelmingly in your debt, if there can be accounts in love. What it has been to me to live this time in your heart and companionship no phrases can convey.*

*Much, much love,*
*Robert*

As I read the words again, my eyes welled in the memory of that time, of our first year, and of all the years since then. But just as Clementine felt toward Winston, I feel it is Robert who enriches my life. I feel indebted to him. Through Robert I have been inspired. Inspired to love more deeply than I ever thought I could. To endure rigorous training I never thought I wanted. To write about things I never knew existed. Because he and I became us, I gained perspectives few can say they embody all at once—as doctor, as patient, as someone who loves a patient. Already we have been through so much together. Some of it has been hard, tearful, and unplanned, but most of it has been more than we knew to hope for and none of it do we regret, because all of it has brought us here together in this moment to experience what lies ahead.

# ACKNOWLEDGMENTS

So many thanks to the patients and families whose stories inspire me to write.

Alita Anderson generously gave her time, energy, and advice to bringing this book to fruition though her own plate was already overflowing. She is a true friend and the most genuinely positive person I know, able to point out how what felt like a pile of crap to me was really the fertilizer I needed to grow a beautiful flower garden.

Delphine Tuot and Elaine Ku were instrumental in helping to make sure my artistry was actually consistent with real nephrology. Lynn Mazur, social worker and my PCEP buddy, showed me a way to engage patients dealing with terminal illness that my medical doctor training did not provide. Melanie Tervalon, who now gladly takes credit for bringing Robert and me together, was a part of the small circle of people whom I trusted with earlier book versions.

I'm thankful to Talmadge King for not only agreeing to be my writing protector, but also being nearly impossible to offend. I'm thankful to all the other family, friends, and colleagues who have been excited for me and have encouraged me throughout this process: Avery Burt, Janet Smith, Karla Jones, Jennifer Gunn, Mark Smith, Alice Chen, Bob Wachter, and everyone who has read my blog, particularly those who let me know what my writing has meant to them.

Chris Freise welcomed me into his operating room and explained the details of kidney transplant surgery. Because of him I was able to write about my and Robert's surgeries in a way that my patient/primary care doctor–turned–nephrologist self would have not otherwise been able to do. Rachel Howard helped me refine my original book proposal. Victoria Sweet believed in my writing ability when I wasn't so sure and introduced me to our agent, Mary Evans. Mary guided the book in a direction that I never intended but am now sure was the very best way to go. Tracy Sherrod, my editor, fought for this book, sometimes even with me. It took us a while to hear and trust each other, but getting there has led to what I believe is a better book than I could have written alone.

Finally, I thank my husband and love of my life, Robert Phillips. He has been my rock and my greatest cheerleader. It was his encouragement that made me start writing to begin with. And without him there would be no book.

# FREQUENTLY ASKED QUESTIONS

***I have pain on the side of my lower back. Is it because of my kidney?***

Unless you've been having pain when you pee for a while (suggesting a bladder infection that has spread up to a kidney) or have been seeing blood when you pee (possibly because of kidney stone), probably not. There are nerves and muscles back there too.

***How can I find out if I have a kidney problem?***

A simple blood test for creatinine (say "cre-at-uh-neen") and urine test for protein or tiny amounts of blood only visible by microscope can detect a kidney problem early. High blood creatinine is a sign your kidneys have been damaged and aren't filtering your blood fast enough. Blood or protein in your urine can mean the kidney filters have been damaged. Have these tests done every year.

If you develop symptoms such as pain when you pee, blood in your pee, problems peeing, pee that bubbles up in the toilet like foam, or swollen legs, go get checked out even if you just had your annual checkup. Blood and pain when you pee could be caused by kidney stones or infection, but lots of other diseases can cause blood in your pee too. Problems peeing (like having to push to start peeing, dribbling after you thought you were finished peeing, or having to get up and pee two or more times overnight) suggest there is something blocking the urine flow, such as the prostate or a large mass. Urine that can't pass backs up into the kidneys, causing pressure that can irreversibly kill kidney cells if left there too long. Foamy pee and swollen legs suggest there is lots of protein in the urine (though non-kidney-related diseases can cause swollen legs too).

***How can I find out what caused my kidney problem?***

While we usually check various blood and urine tests and look at urine with a microscope to get a sense of why someone has blood or protein in their urine, often a kidney biopsy is necessary to know exactly why—especially if the blood cells in the urine are misshapen or if there is 1 gram (1,000 milligrams) or more of protein in the urine. More than a few blood cells in the urine is never normal, but when they are shaped funny there is a problem with the kidney filters (glomeruli). Knowing exactly why allows more targeted treatment, helps us see how much irreversible damage was done, and helps us better predict prognosis. See chapter 12, Zebras, for more details on what a kidney biopsy entails.

***How can I know how serious my chronic kidney disease is?***

We estimate kidney function with an equation that takes into account the person's blood creatinine concentration, age, gender, and, unfortunately, race (if Black or not) as a proxy for muscle mass. The result is a measure of how fast blood is being filtered through the kidneys, the estimated glomerular filtration rate (eGFR). In a young, healthy adult, the eGFR can be as high as 125 milliliters per minute—roughly 1/2 cup or 25 teaspoons of blood being filtered through the kidneys every minute of every day. Above 90 milliliters (18 teaspoons) per minute is considered normal kidney function. After age forty, we lose about 1 milliliter per minute every year (or 1 teaspoon every five years). We classify kidney disease into five main stages:

| Stage | Estimated GFR (milliliters/ minute) | Equivalent kidney function in teaspoons | Level of CKD* | Description |
|---|---|---|---|---|
| 1 | 90+ | 18 | At risk | Normal kidney function but has a risk factor for kidney disease (like a genetic trait or family history of diabetes) |
| 2 | 60-89 | 14-not quite 18 | Mild | Mildly decreased kidney function and abnormal amounts of blood or protein in the urine |
| 3 | 30-59 | 6-not quite 14 | Moderate | Start to see signs that kidneys aren't working properly (like anemia or high phosphorus) |

| Stage | Estimated GFR (milliliters/minute) | Equivalent kidney function in teaspoons | Level of CKD* | Description |
|---|---|---|---|---|
| 4 | 15–29 | 3–not quite 6 | Severe | More problems with anemia, acid, calcium, and phosphorus |
| 5 | <15 | <3 | End-stage kidney disease | Replacement usually needed about 5–8 ml/min (1–1 ½ teaspoons) |

**CKD = chronic kidney disease*

The more albumin (the main type of protein) a person has in their urine at any stage, the more at risk they are of worsening chronic kidney disease, sudden temporary kidney damage (acute kidney injury), death from heart disease, and death from any cause. Urine albumin higher than 300 milligrams is more risky than urine albumin between 30 and 300 milligrams. Urine albumin less than 30 milligrams is normal.

***How do I prevent chronic kidney disease?***

Fortunately, most causes of kidney disease can be prevented. For every three people with chronic kidney disease in the United States, diabetes and high blood pressure are the causes of two of them. While diabetes and high blood pressure tend to run in families, one can lower their own risk by not putting on too much weight, by not drinking too much alcohol, by not eating too much salt, and by not smoking.

Other ways of preventing chronic kidney disease: Don't use

cocaine or heroin. Don't take nonsteroidal anti-inflammatory drugs (NSAIDs) like ibuprofen, Motrin, Aleve, naproxen, Naprosyn, or aspirin (except a baby aspirin for heart protection) regularly for more than a few days. Don't have unprotected sex with someone who has HIV or hepatitis B. Don't share a toothbrush or razor with someone who has hepatitis C.

***I heard a person could live with just one kidney. How much is each of my kidneys affected?***

Unless you have something that is in just one place, such as a blockage of the large artery that funnels blood into the kidney or a big mass or stone, both kidneys are equally affected. Both kidneys are feeling diseases (like diabetes or high blood pressure) that affect the entire body (systemic diseases).

***Does alcohol hurt my kidneys?***

The liver, not the kidneys, processes alcohol, which is why alcoholics sometimes suffer from liver failure. However, alcohol can *indirectly* hurt the kidneys if one drinks enough often enough to cause high blood pressure, which can directly cause kidney damage.

***Why is there a warning on my deodorant bottle to not use it if you have kidney disease?***

Some deodorants contain aluminum, very tiny amounts of which can *theoretically* be absorbed through the skin into the blood. Toxic levels of aluminum can cause a weakening of bones, anemia (low blood levels) resistant to treatment with iron, and dementia. This used to be a common problem in patients with advanced chronic kidney disease or on hemodialysis until the 1980s when the main sources of aluminum—medications to lower blood phosphorus levels and municipal water sources—were dealt with. Now that aluminum-based binders are rarely used in the United States and strict national and international guidelines for removing aluminum from water supplied to dialysis centers are in place, aluminum toxicity is essentially a thing of the past.

***What do I need to eat or drink to make my kidney function come back or keep my kidneys from getting worse?***

Unfortunately, there is no food or drink that can fix kidneys. At best, one can slow down the pace at which kidneys worsen by eating a low-carbohydrate diet to help control diabetes or a low-salt diet to help control high blood pressure. A low-protein diet can slow down worsening somewhat in people who lose a lot of protein in their urine, but this often comes at the expense of the person's overall body nutrition and muscle health.

A kidney or renal diet is about restricting the amount of potassium or phosphorus when the kidneys can no longer control them. Following a low-potassium diet can protect you from dangerously high blood potassium levels that can stop your heart. Following a low-phosphorus diet can help maintain strong bones. But a renal diet does not make the kidneys better or slow down the progression of the disease. And since it involves limiting a lot of really good food, there is no need to follow it unless the person is at a very advanced or end-stage (on dialysis) kidney disease.

***How much water should I drink to flush out my kidneys?***

While it is important to not allow yourself to become dehydrated, your blood is constantly being filtered through working kidneys.

That's how they remove toxins from the blood. There is no need to flush them. Drink if you're thirsty.

***What medicine do I need to make my kidney function come back or keep it from getting worse?***

Unfortunately, with chronic kidney disease, damage to the kidneys is irreversible. One can slow down the rate at which it worsens by controlling things that can hurt the kidneys (like diabetes or high blood pressure) or by avoiding things that can hurt the kidneys (like NSAIDs or too-high doses of certain antibiotics). Certain diseases can be treated with medicines that suppress the immune system. Angiotensin converting enzyme (ACE) inhibitors can slow down worsening by lowering the protein in the urine.

***Can cranberry juice protect my kidneys?***

Cranberry juice is a weak antibiotic, which means it can partially treat a bladder infection. The problem with treating a bladder infection partially is that it can mask the symptoms (the pain when you pee may lessen) while the infection spreads to the kidneys. Don't rely on it to adequately treat an infection or to prevent anything. It can't.

***If I have chronic kidney disease, how soon will I need dialysis?***

It depends on the cause, how early it's detected, and how it is treated. Some diseases (particularly those that cause more than 2,000 milligrams of protein in the urine) can cause the kidneys to completely fail within five or ten years. Some causes take so long to worsen to complete kidney failure that the person will die of something else first. Repeated blood and urine tests are critical to getting a better sense of the rate at which your kidneys are declining.

***I feel fine and pee fine. Why do I need dialysis?***

All pee ain't good pee. The kidneys do so much more than just pass water. Urine must also contain enough potassium, acid, and toxins for the body to survive.

Most people do feel pretty well until it's time to start dialysis. When that is depends on the size of the person, because the bigger the person, the more kidney they need. But most people will need dialysis when the estimated glomerular filtration rate (eGFR, how fast blood is being filtered through the kidneys) drops down to near 5 milliliters (1 teaspoon) per minute. But more important than the eGFR is how the person is doing (if there is nausea, vomiting, unintentional weight loss, sleeping more during the day, shortness of breath) or if the potassium level, acid level, or blood pressure can't be controlled with medications.

***What does dialysis do and how will it make me feel?***

Dialysis is a kidney replacement; it does many of the jobs the kidneys can no longer do, like removing toxins and excess water, potassium, acid, and phosphorus and replacing calcium. Dialysis doesn't work nearly as well as normal kidneys do. It really only gives a person the equivalent of severe chronic kidney disease, an eGFR of 20 to 25 milliliters (4 to 5 teaspoons) per minute. And it doesn't correct the anemia or fully control the phosphorus. Medications can be given to help with these.

Patients on hemodialysis (through the blood) need to follow a low-potassium diet and limit their fluid intake between dialysis treatments (three times a week) because potassium and fluid can only leave the body through dialysis. Most hemodialysis patients stop urinating much at all within several months after starting dialysis. They may feel if their potassium level becomes dangerously high (though most don't feel high potassium at all) or feel short of breath or have increased leg swelling if they drink too much fluid between treatments. Patients on hemodialysis may feel tired after their treatment. How long depends on their age and their overall health, from a couple of hours to until it's almost time to go back to dialysis. During dialysis, some get cramping or light-headed if too much fluid is removed. People can also have problems like infection (especially with hemodialysis catheters) and clotting (especially with grafts; more than with fistulas). Grafts are more prone to clotting than fistulas, because a graft is a manmade blood vessel, while a fistula is made from the person's own blood vessels.

Patients on peritoneal dialysis (through the belly) tend to feel more "even" and don't have as much fluid restriction and often no potassium restriction since dialysis is being done continually every day. They don't get cramping or light-headedness, but they may get an infection (peritonitis) if they don't take good care of their peritoneal catheter site and have excellent technique (wash hands, wear a mask, and be careful when draining or filling the belly). Peritonitis is uncommon and usually can be treated with two to three weeks of antibiotics at home.

***If I start dialysis, how long will I need it?***

Until you get a transplant. Dialysis is just a kidney replacement. It's not making the kidneys better.

***Why do I have to have dialysis? Can't I just get a kidney transplant?***

Maybe. If you are healthy enough for the transplant surgery and someone is willing to give you one of their kidneys or you are placed on the kidney waiting list when your eGFR is 20 milliliters per minute and your disease worsens slowly, you may get a transplant without ever having to be on dialysis. Unfortunately, many people are not aware they have chronic kidney disease or are not in the care of a nephrologist, so referral for kidney transplant can be delayed. Further, most kidneys for transplant come from deceased donors and many more people need kidney transplants than kidneys are available, so most people will need to spend some time on dialysis before a kidney transplant is available. Finally, paying someone for their kidney is considered unethical.

***If I have a kidney transplant, where does it go and do they take out my old kidneys?***

Usually the transplanted kidney will be placed in the lower abdomen on the right side. It takes a couple of hours to do the kidney transplant. Taking out the original (native) kidneys would take another three to five hours, so they usually are not removed.

However, the native kidneys may be removed in these cases:

- Patients with polycystic kidney disease whose kidneys cause a lot of pain, blood in urine, or infection, or are so big there's no room for the new kidney
- Patients who have a lot of infections because abnormal anatomy cause urine to go the wrong way (reflux)

- Patients who are losing very high amounts of protein in their urine (nephrotic range proteinuria, more than 3 grams a day)
- Patients who have a small mass in their kidney that might be caused by cancer

***Can I travel if I am on dialysis? Work? Exercise?***

Yes. I encourage people to work and exercise as much as they feel up to. Arrangements can be made for hemodialysis patients to visit a dialysis center where they are going. Peritoneal dialysis supplies can be shipped. However, travel outside of your state or to some countries may not be possible if you don't have Medicare or private health insurance or if hemodialysis centers are not available.

***When should someone consider not starting dialysis or stopping it?***

Dialysis significantly lengthens and improves the quality of life for most people with end-stage kidney disease. However, dialysis may not extend life and may even worsen the quality of life for people who are over age seventy-five and have other serious medical problems such as dementia or ischemic heart disease (blockage of heart blood vessels). Dialysis may prove to be more of a burden. This group may be better cared for with conservative management—which includes symptom management and psychosocial/spiritual support. These patients may live weeks, a few months, or even a couple of years depending on how long their kidney function lasts. In the end, they will become sleepier and sleepier over time. Some may develop nausea, vomiting, or cramping. One patient I cared for had a seizure. There are medications to minimize all of these symptoms.

Stopping or withdrawing dialysis should be considered when things change for the worse. For example, if a person is diagnosed with a new terminal condition or has significant worsening of an existing condition, then stopping should be considered, particularly if dialysis is becoming harder to do and doesn't seem to be helping the person feel any better. People who have very little or no kidney function and withdraw from dialysis completely will die within a few days or up to two weeks.

***How can I support someone who has chronic kidney disease, who is on dialysis, or has a kidney transplant?***

Different people have different needs depending on where they are in the course of their illness. And different people have different personalities with varying abilities to ask for or accept support. Some will fear that they are or will be a burden to those around them. Ask. Offer.

For someone just finding out about their diagnosis or that they may soon need dialysis, offer to go with them to see the doctor or to educational sessions. It's difficult for some people to hear, comprehend, *and* think of what questions to ask—especially when the news is upsetting. Encourage them to write down questions before appointments. Take notes on what is said.

If you live with someone who has very advanced chronic kidney disease or is on dialysis and needs to follow a low-potassium and/or low-phosphorus diet, understand what their restrictions are and be supportive. For example, if the person needs to follow a low-phosphorus diet, don't bring liters of Diet Coke or Pepsi home; Diet Sprite and 7UP are preferable for soda drinkers.

Someone who is about to have a kidney transplant will need support in those first few months. They will need rides to and from doctor appointments or the lab. They may need help changing bandages or picking up medications from the pharmacy.

Wherever a person may be in the course of their illness, it is important for the people who care for them to have empathy, not pity, for their situation. To support their self-sufficiency, not mother them. And, most important, to just be there and not run.

Still more questions? Check out the National Kidney Foundation (www.kidney.org) or American Association for Kidney Patients (www.aakp.org) for additional information.

## ABOUT THE AUTHOR

**Vanessa Grubbs, MD,** is an associate professor of medicine and nephrology at the University of California, San Francisco, and maintains a clinical practice and research program at Zuckerberg San Francisco General Hospital. She received her undergraduate and medical degrees from Duke University and teaches writing for patient advocacy to medical students and practicing physicians. She lives with her husband, teenage son, and two dogs in Oakland, California. This is her first book.